T0419254

IMMUNOLOGY AND IMMUNE SYSTEM DISORDERS

# DNA VACCINES: TYPES, ADVANTAGES AND LIMITATIONS

# IMMUNOLOGY AND IMMUNE SYSTEM DISORDERS

Additional books in this series can be found on Nova's website under the Series tab.

Additional E-books in this series can be found on Nova's website under the E-books tab

IMMUNOLOGY AND IMMUNE SYSTEM DISORDERS

# DNA VACCINES: TYPES, ADVANTAGES AND LIMITATIONS


ERIN C. DONNELLY
AND
ARTHUR M. DIXON
EDITORS

Nova Biomedical Books
*New York*

**Library of Congress Cataloging-in-Publication Data**

DNA vaccines : types, advantages, and limitations / editors, Erin C. Donnelly and Arthur M. Dixon.
p. ; cm.
Includes bibliographical references and index.
ISBN 978-1-61324-444-9 (hardcover : alk. paper) 1. DNA vaccines. I. Donnelly, Erin C. II. Dixon, Arthur M.
[DNLM: 1. Vaccines, DNA. QW 805]
QR189.5.D53D64 2011
615'.372--dc22

2011012604

*Published by Nova Science Publishers, Inc. † New York*

# Contents

# Preface

DNA vaccination is one of the most promising techniques for immunization against diseases caused by viruses, protozoa, bacteria, and even for tumors and illnesses with genetic origins. These vaccines can be administered by direct inoculation of plasmid by several routes. In this book, the authors present topical research in the study of the types, advantages and limitations of DNA vaccines. Topics discussed include the role of novel carrier constructs in effective and safe DNA vaccine delivery; improving the immunogenicity of DNA vaccines; the role of DNA-IL-12 vaccination in cosinphilic inflammation; genetic immunization enabling re-direction of the host immune system resulting in the development of effective immune responses; and DNA vaccination against herpesvirus.

Chapter 1 – The DNA based vaccines are typically consisted of a DNA sequence that encodes for an antigenic protein in vivo. The fragments of antigenic peptide are then MHC I restricted presented onto the surface of APCs or transfected cells directing generation of cell mediated and humoral immune responses with simultaneous generation of immune memory. Thus on subsequent exposure protection against the invading pathogen is conferred. Being most safe, reproducible and easier to produce, DNA vaccines attract immense attention, however they are poorly immunogenic due to their poor cellular uptake, short half-life and low biological stability. Additionally, biological barriers also heavily hamper the cellular delivery of these molecules. Numerous efforts have been made for successful delivery of these molecules and a number of physical and chemical techniques have been developed, optimized and evaluated for the purpose. The novel carriers including liposomes, polymeric particles, dendrimers etc. have gained significant attention and success in the safe and effective delivery of DNA. The present chapter discusses the role of novel carrier construct(s) in effective and safe DNA vaccine delivery.

Chapter 2 -A major biotechnological revolution of the last two decades has resulted in the novel application of DNA as a vaccine to prevent or treat disease. This has progressed from principle to product in only 20 years. However, the wider application of DNA vaccines is reliant on improving their immunogenicity, as while DNA vaccination can provide a broad immune response to an antigen, the magnitude is often less than optimal. This has led to many studies into improving the immunogenicity by using different delivery methods, targeting to specific cells in the host and by co-delivery with potent immune stimulators. This chapter reviews the different methodologies used to improve the immunogenicity of DNA vaccines, designed to maximise immunogenicity without changing the fundamental simplicity and usability of this type of vaccine. The most recent innovations, of delivery using nanoparticles, are also highlighted.

Chapter 3 -DNA vaccination has been dubbed the "third revolution" in vaccine development by some observers ever since its promising in vivo application in the early 1990s. The method is an attractive vaccine avenue because of its inherent simplicity as well as its relative cost effectiveness. DNA vaccination is also appealing because it is possible to produce large quantities of vaccine in a short period of time. In comparison with other conventional vaccines, DNA vaccines are stable and have a long shelf life. The most intriguing property of a DNA vaccine is that it can promote long-lasting humoral and cellular immunity leading to protective and/or therapeutic effects in preclinical and clinical tests. However, the immunogenicity of a DNA vaccine is hindered by suboptimal delivery of the DNA to cells, especially in large animals. Several other concerns have been raised in the field including the potential for DNA integration into the host genome and the possibility of autoimmune responses due to the long-term presence of the foreign DNA. In this review, different approaches to optimize DNA vaccines in order to improve their immunogenicity and delivery are presented. Discussion will include the choice of the antigen, codon optimization to increase antigen expression, choice of optimal expression vectors and proper adjuvants; there will be a special focus on several novel delivery methods. Also, the potential strategies to address the concerns in DNA vaccine application.

Chapter 4 -Traditional vaccine approaches have proven to be insufficient against a variety of diseases, mainly those without spontaneous cure. In the last years, recombinant DNA technology has emerged as a promising tool for vaccine development against infectious agents, cancer, autoimmunity and allergy. DNA vaccines are based on the delivery of genes encoding the antigen of interest that can be translated by host cells. Importantly, both humoral and cell-mediated immune responses may be elicited against multiple defined antigens simultaneously. It offers several potential advantages over conventional approaches, including safety profile and feasible production method. Despite these advantages, the limited immunogenicity of DNA vaccines in humans has hampered their use in the clinical setting. Recently, four DNA vaccines have been approved for veterinary use, suggesting in a near future their employment in humans. Further optimization of DNA vaccine technology, including rational plasmid design, different delivery systems, addition of adjuvants, mainly biological modifiers like some bacteria or their compounds, and improved immunization protocols is an interesting alternative to achieve the immunogenic properties that have been demonstrated in preclinical models. This chapter will discuss/focus on the main features of DNA vaccines.

Chapter 5 -Background and Aims: DNA-based vaccination is a promising novel approach for immunotherapy of chronic Hepatitis B Virus (HBV) infection, however its efficacy needs to be improved. In this preclinical study are the evaluated, therapeutic benefit of cytokine (IL-2, IFN-$\gamma$) genes co-delivery with DNA vaccine targeting viral proteins in the duck model of chronic HBV (DHBV) infection. DHBV is an avian hepadnavirus with replication strategies and genomic organization closely related to the human virus. Experimental DHBV inoculation of neonates always leads to the establishment of a chronic DHBV-carrier state, representing therefore a reference model for evaluation of novel therapeutic approaches on viral clearance, particularly on covalently closed circular (ccc) DNA elimination, viral minichromosome that is responsible for the chronicity of infection.

A colony of chronically infected Pekin ducks was established and treated by DNA vaccine encoding DHBV structural proteins (envelope, core) alone or co-delivered with either duck IL-2 (DuIL-2) or duck IFN-gamma (DuIFN-γ) gene. The impact of such therapy on viral replication was followed by monitoring of viremia, break of immune tolerance and intrahepatic viral liver DNA analysis.

*Results and Conclusion:* Shows that co-administration of cytokines genes (IL-2, IFN-γ) with DHBV DNA vaccine may represent a promising strategy able to enhance viral clearance from serum and liver of chronically infected carriers. IFN-γ gene co-delivery showed a particularly effective adjuvant activity able to enhance therapeutic efficacy of DNA vaccine targeting hepadnavirus proteins, leading to drastic inhibition or suppression of viral replication in about 50% of animals, which have only traces of cccDNA. Interestingly, the inhibition of viral replication was sustained, since no rebound of viral replication was observed during three months after immunotherapy cessation. However, some livers of apparently resolved animals were able to transmit DHBV infection to neonatal ducklings. No adverse effect was associated with co-imunnization with cytokine genes.

Collectively our results suggest a therapeutic benefit of DuIFN-γ gene co-delivery with DNA vaccine targeting viral proteins (envelope, core) for chronic hepatits B therapy. The efficacy in term of viral clearance was higher as compared with previous studies in which the therapeutic efficacy of DHBV DNA vaccine in combination or not with antiviral drugs (adefovir, lamivudine) has not exceeded 30%.

Chapter 6 -DNA-based vaccines have garnered attention for their potential as alternative treatments for established diseases. Eosinophilia is a key inflammatory feature of allergic pathologies including asthma. This pathology is characterized by airway obstruction, airway hyper-responsiveness and lung tissue remodeling. IL-12 is a potent inducer of Th1 cellular immune responses, but maintenance of the Th1 immune response is necessary for increased IFN-γ production. Th1 cytokines such IL-12 and IFN-γ have been shown to reduce lung allergic responses. Previously reported is that DNA-IL-12 vaccination leads to a persistent reduction in blood/bronchoalveolar eosinophilia following *Toxocara canis* infection. Prominent type-1 immune response has been identified as the hallmark of *T. canis* infection following DNA-IL-12 vaccination. The type-1 polarized immune-logical profile in DNA-IL-12-vaccinated animals is characterized by a high IFN-γ/IL-4 ratio and low levels of IgG1, with subsequent high IgG2a/IgG1 ratio. The persistent airway hyper-responsiveness observed in DNA-IL-12-vaccinated animals demonstrates that the airway constriction observed involves immunological mediators other than those blocked by DNA-IL-12. The data discussed here provide evidence of the potential utility of DNA-IL12 vaccine for the development of a new treatment for eosinophilic inflammation. The purpose of this study is to review DNA vaccination, a novel strategy of DNA administration, as well as the role of DNA-IL-12 vaccination in eosinophilic inflammation.

Chapter 7 -Historically, immunization has been designed to prevent the onset of infectious disease, with vaccines acting as prophylactic agents. However, over the course of time immunization has evolved to include a therapeutic objective whereby individuals who have already contracted an infectious or neoplastic disease, and where traditional treatment options are limited, are vaccinated. With diseases such as HIV and tuberculosis the limited success of traditional chemotherapeutic and prophylactic approaches has resulted in the experimental advancement of therapeutic genetic immunization, where the host immune system is to be modulated to improve disease prognosis or ultimately, eradicate the infection. Furthermore, for cancer, a major non-infectious disease, similar experimental treatments are underway.

Chapter 8 -Infectious diseases inflict heavy economic losses in cattle, although they are primarily preventable by vaccination. Among the different types of vaccines available, DNA immunization is one of the most recent and technologically advanced strategies to immunize against a highly prevalent infectious agent: bovine herpesvirus type 1 (BoHV-1). In the last ten years, DNA technology has offered the opportunity to retain the advantage of recombinant expression of immunodominant antigens while overcoming safety issues and logistical inconveniences inherent to live replicating vectors. Different studies on DNA vaccines against BoHV-1 have been based on expression of gD gene with a truncated trans-membrane domain. The immunizing DNA can be inoculated by conventional route (syringes) or by gene guns and with electroporation system, in order to increase the humoral and cell mediated immune responses. Different adjuvants have been inserted in DNA vaccines. However, the mechanism of action of these adjuvants is not yet completely understood and it is known that different adjuvants use different ways of estimulating the immune system.

The expressed proteins can readily go through processing and presentation via class II and class I pathways, which will result in induction of CD4+ and CD8+ T-cell responses. DNA vaccines present, howler, several inconveniences, for instance, they can stimulate immunity exclusively against proteins but not against polysaccharides; thus, at present time there is not valid practical option for polysaccharide-based-vaccines. Safety is a strong qulity of DNA vaccines: as there is no complete homology between mammalian cells and plasmid DNA occurrence of a homologous recombination is highly unlikely.

Chapter 9 -DNA vaccination is one of the most promising techniques for immunization against diseases caused by viruses, protozoa, bacteria, and even for tumours and illnesses with genetic origins. These vaccines can be administered by direct inoculation of plasmid by several routes. The DNA immunization can induce both humoral and cellular immune responses, directed to the CD4 helper T cells and CD8 cytotoxic T responses, production and presentation of antigen to the immune system in a manner similar to that of a natural infection, enabling lasting immune responses and the possibility of combining a few antigens for the simultaneous treatment of many diseases. Another major advantage of DNA vaccines is the low toxicity, as DNA is completely innocuous. The use of DNA vaccines offers a number of economic, technical and logistics, advantages as compared to classical vaccines. Its mass production is cheaper, allows for easier control of the quality, and marketing does not require a cold chain, because these vaccines are stable at room temperature and endure pH variations. These factors facilitate the transport and distribution, and allow the transfer of technology. Despite the potential of DNA vaccines is not always fully achieved, since it depends on the nature of antigens, frequency and route of administration, the concentration of DNA administered targets the cellular localization of the antigen encoded by plasmid regardless of host age and health. Regarding the security of this new technology, one must take into account the possibility of plasmid integration into the host genome as potentially damaging, as exposure to high levels of antigenic proteins of the immune system during the period of gene expression may lead to generation of autoimmune diseases. In short, although this methodology still has some way to go before it can be safely administered in humans or animals to cure disease; this method allows the development of scientifically progress and maturity.

In: DNA Vaccines: Types, Advantages and Limitations ISBN 978-1-61324-444-9
Editors: E. C. Donnelly and A. M. Dixon 

*Chapter I*

# Novel Carrier Adjuvant for DNA Vaccination

***Suresh P. Vyas,[1] Bhuvaneshwar Vaidya and Shailja Tiwari***
Drug Delivery Research Laboratory, Department of Pharmaceutical Sciences
H. S. Gour University, Sagar (M.P.) 470003, India

## Abstract

The DNA based vaccines are typically consisted of a DNA sequence that encodes for an antigenic protein in vivo. The fragments of antigenic peptide are then MHC I restricted presented onto the surface of APCs or transfected cells directing generation of cell mediated and humoral immune responses with simultaneous generation of immune memory. Thus on subsequent exposure protection against the invading pathogen is conferred. Being most safe, reproducible and easier to produce, DNA vaccines attract immense attention, however they are poorly immunogenic due to their poor cellular uptake, short half-life and low biological stability. Additionally, biological barriers also heavily hamper the cellular delivery of these molecules. Numerous efforts have been made for successful delivery of these molecules and a number of physical and chemical techniques have been developed, optimized and evaluated for the purpose. The novel carriers including liposomes, polymeric particles, dendrimers etc. have gained significant attention and success in the safe and effective delivery of DNA. The present chapter discusses the role of novel carrier construct(s) in effective and safe DNA vaccine delivery.

---

1 Email: vyas_sp@rediffmail.com.

# 1. Introduction

Vaccines are considered as the major therapeutic strategy as they offer effective and valuable prevention against infectious diseases. Vaccines are administered prophylactically with the objective to elicit long term immunological protection and generation of immune memory, therefore the first encounter of pathogen is 'remembered' by the immune system. The conventional mode of vaccination includes use of live or attenuated whole pathogen for immunization. However, the use of pathogen is associated with a number of drawbacks as reversal of virulence and possibilities of generation if hypersensitive reactions. Further, the process of empirical attenuation is unreliable in the case of highly virulent pathogenic organisms such as HIV, influenza etc.

As an alternative to live pathogen vaccines recombinant protein antigen and DNA based vaccines are developed. These vaccines are based on the rational design of vaccines with identifying immune correlates of protection as a response against the recombinant protein (protein antigen vaccines) or designing of DNA that expresses the protein *in vivo* following administration. Therefore, the important step in a rational design of a vaccine is to understand the form a mechanistic perspective so that the designed vaccines activate and expand memory B and T cells, which are then poised to respond rapidly and specifically to the subsequent exposure to the pathogen.

DNA Vaccines have been one of the latest developments in vaccine technology. The DNA based vaccines are dependent upon high molecular weight double stranded DNA construct containing transgene which encodes specific proteins. The plasmid molecules gain access in to the nucleus from cytoplasmic release and there are transcribed into RNA and finally translated into desired protein antigen.

The concept of DNA vaccination was very first established in 1990, by Wolff *et al.* who demonstrated that the direct injection of naked plasmid DNA (pDNA) encoding a foreign antigen into mouse myocytes resulted in uptake of the pDNA by host cells and subsequent expression of the foreign antigen (Wolff *et al.*, 1990). Subsequently, in a number of animal studies it was clearly demonstrated that pDNA vaccines were capable of eliciting humoral (Tagawa *et al.*, 2003) and cellular (Ulmer *et al.*, 1993, Sedegah *et al.*, 1994) immune responses against the encoded vaccine antigen. Following these pioneering studies the pDNA vaccines have been tested to be effective in eliciting HIV antigen-specific immune responses in a wide variety of animal model systems (Babiuk *et al.*, 1999; Donnelly *et al.*, 1997; Egan *et al.*, 2000) and also have been evaluated in a number of phase I clinical trials (few of them are summarized in table 1), where they have been found to be well tolerated.

**Table 1. Phase I clinical trial with DNA based vaccines**

| DNA based vaccine approach | Disease investigated | Outcome of the investigation | Reference |
|---|---|---|---|
| DNA-based vaccine containing HIV-1 Env and Rev genes was tested | HIV seropositive patients | Enhanced specific lymphocyte proliferative activity against HIV-1 envelope was observed in multiple patients | Boyer *et al.*, 1999; 2000 |
| DNA based vaccine encoding Carcinoembryonic Antigen and Hepatitis B Surface | For hepatitis B and carcinoembryonic antigen (CEA) | Repetitive dosing of the DNA vaccine induced HBsAg antibodies in 6 of 8 | Conry *et al.*, 2002 |

| DNA based vaccine approach | Disease investigated | Outcome of the investigation | Reference |
|---|---|---|---|
| Antigen | vaccination in Colorectal Carcinoma Patients | patients, with protective antibody levels achieved in four of these patients. CEA-specific antibody responses were not observed, but 4 of 17 patients developed lymphoproliferative responses to CEA after vaccination. | |
| Heterologous prime boost immunization study that includes priming with PfCSP encoding DNA vaccine and boosting was done with recombinant protein antigen vaccines | *Plasmodium falciparum* malaria vaccination | An antibody response as well as enhanced T cell response was observed in the patients in the immunized individuals as compared to protein or DNA alone. | Epstein *et al.*, 2004 |
| Intramuscular injections of 1 mg of a DNA vaccine encoding HBV envelope proteins | *Hepatitis B vaccination* | Proliferative responses to hepatitis B surface antigen with concomitant increase in HBV-specific interferon gamma-secreting T cells were detected as measured by proliferation, ELISApot assays, and tetramer staining. Secondary end points included safety and the monitoring of HBV viraemia and serological markers. | Mancini-Bourgine *et al.*, 2004 |
| DNA encoding a chimeric immunoglobulin molecule containing variable heavy and light chain immunoglobulin sequences derived from each patient's tumor, linked to the IgG2a and κ mouse immunoglobulin (MsIg) heavy-and light-chain constant regions chains, respectively | *Patients with B-Cell Lymphoma* | 7 of 12 patients mounted either humoral (n=4) or T-cell-proliferative (n=4) responses to the MsIg component of the vaccine and in one patient, a T-cell response specific to autologous Id was also measured | Timmerman *et al.*, 2002 |
| A novel DNA based Hepatitis B vaccine administered intraepidermally by particle mediated epidermal delivery | Hepatits B vaccine | The protective immune response was generated even in the subjects who suboptimally responded to conventional vaccination | Rottinghaus *et al.*, 2005 |
| DNA plasmid encoding tyrosinase epitopes | vaccine for patients with Stage IV melanoma cancer | Immune responses by peptide-tetramer assay to tyrosinase 207-216 were detected in 11 of 26 patients. Further, Survival of the heavily pretreated patients on this trial was unexpectedly long, with 16 of 26 patients alive at a median follow-up of 12 months | Tagawa *et al.*, 2003 |

**Table 1. (Continued)**

| | | | |
|---|---|---|---|
| Plasmid DNA vaccine encoding H5 hemagglutinin-encoding plasmid | influenza A virus vaccine | hemagglutination inhibition (HI) titers of > or =40 and 4-fold rises from baseline were achieved in 47-67% of subjects and H5-specific T-cell responses was elicited in in 75-100% individuals. | Smith *et al.*, 2010 |
| *Other recent developments* | | | |

| Company | Trial with DNA based vaccine | Doi |
|---|---|---|
| Vical Pharma | Begins Phase 1 Trial of DNA Vaccine Against H1N1 Pandemic Influenza | http://ir.vical.com/releasedetail.cfm?ReleaseID=466603 |
| *Inovio Pharmaceuticals'* | Begins phase I clinical trial with cancer (VGX-3100), Avian influenza (VGX-3400X), HIV preventive vaccines (PENNVAX$^{TM}$-G), HIV therapeutic vaccine (PENNVAX$^{TM}$-B), Hapatitis C virus NS3/4A | http://www.inovio.com/products/index.htm |
| *Vaxono* | DNA based vaccine against malaria | http://www.vaxono.com/ |
| *PowderMed* | PowderMed Initiates a Phase I Clinical Trial for a Novel Therapeutic Herpes Simplex Type 2 (HSV2) DNA Vaccine | http://www.powdermed.com/ |
| *Lipotek Pty Ltd* | Early phase clinical trial of a melanoma immunotherapy based on the ***Lipovaxin*** | http://www.lipotek.com.au/technology/Lipotek_tecknologies. |

The vaccines consisting of naked plasmid DNA have several potential advantages over other alternative immunization approaches such as the delivery of purified or recombinant proteins, or live attenuated or recombinant viruses. They offer the promise of a readily deliverable, molecularly defined reagent that results in antigen synthesis in situ, but that is neither infectious nor capable of replication. Importantly, both humoral and cell-mediated immune responses may be elicited against multiple defined antigens simultaneously.

Also depending upon the selected DNA plasmid construct a long term expression of DNA molecule can be achieved that would eliminate the need of repeated vaccine administration (Boosting). Furthermore, it may become possible to manipulate the nature of the resulting immune response on co-administration of genes encoding immunomodulating cytokines or co-stimulatory molecules (Such as CpG-ODN, cytokine etc.). The genetic constructs can also be modified to allow the removal or insertion of transmembrane domains, signal sequences, or other residues that may affect the intracellular trafficking and subsequently processing of antigen. The sequence may also be modified by site directed mutagenesis, permitting single amino-acid exchanges designed to enhance the antigenic potency of individual epitopes or to abolish unwanted physiologic effects of the wild-type protein. Importantly these plasmid DNA vaccines, encoding suitable antigens or immune modulators, can readily and economically be constructed and produced in large quantities with a high degree of purity and stability.

However, the DNA based vaccination is associated with certain disadvantages too. The comparative advantages as well as disadvantages of DNA based vaccines are presented in Table 2.

# 2. Immunology of DNA Vaccines: Mode of Action and Immune Response Induction

A number of studies suggesting the potential for DNA vaccines revealed that the administration of plasmid DNA *via* various routes and methods elicits antigen specific immune response in experimental animal model and human beings (Donnelly *et al.*, 1997).

**Table 2. Advantages and disadvantages of DNA based vaccines**

| Advantages | Disadvantages |
|---|---|
| • The vaccination with no risk for infection<br>• Presentation of antigen by both MHC class I and class II molecules and can polarise T-cell help toward type 1 or type 2 for stimulation of B cells and CD8+ T cells.<br>• Ease of development and production and stability of vaccine for storage and shipping<br>• Long-term persistence of immunogen<br>• *In vivo* expression ensures that protein more closely resembles normal eukaryotic structure, with accompanying post-translational modifications | • Limited to protein immunogens (not useful for non-protein based antigens such as bacterial polysaccharides)<br>• Risk of affecting genes controlling cell growth<br>• Possibility of inducing antibody production against DNA<br>• Possibility of tolerance to the antigen (protein) produced<br>• Potential for atypical processing of bacterial and parasite proteins |

Furthermore, the investigations demonstrated that the antigen specific antibodies can be detected in rodents for longer than 1 year after DNA immunization (Pardoll *et al.*, 1995). The advantages of using DNA exclude the direct the synthesis of multiple copies of mRNA, hence an expected amplification of both, i.e. the antigen synthesis and corresponding immune response (Leitner *et al.*, 2000).

The understanding of molecular nature of antigen processing, presentation and recognition has shed a light on the precise mechanism of action of DNA based vaccines. The DNA based vaccination basically mimics the natural infection process which triggers both cellular and humoral immune response without any safety related hazards. Following expression the antigenic protein is processed by following two pathways (a) Endogenous: Antigenic protein is presented by the cells in which it is produced. (b) Exogenous: Antigenic proteins is formed within a cell but presented by different cell (Wolff *et al.*, 1990). The findings revealed that the fate of DNA based vaccines is determined by its route of administration and cells which uptake it. If DNA is taken up by an antigen presenting cell (i.e. dendritic cell or macrophages) it will present the antigen via both of the pathways. Alternatively, if it is any other cell the expressed antigen may be presented possibly involving two ways in case the antigen is expressed in the cytosol it is presented to the cell surface in association with MHC-I and also following expression of protein it may be secreted in to the extracellular space from there it is further endocytosed by the APCs processed and presented in association with the MHC-II for induction of humoral immunity.

The kind of immune response generated depends upon the pathway of antigen processing and its presentation (Figure 1) in association with the major histocomptibility complexes (either-I or II) as noncovalent complex.

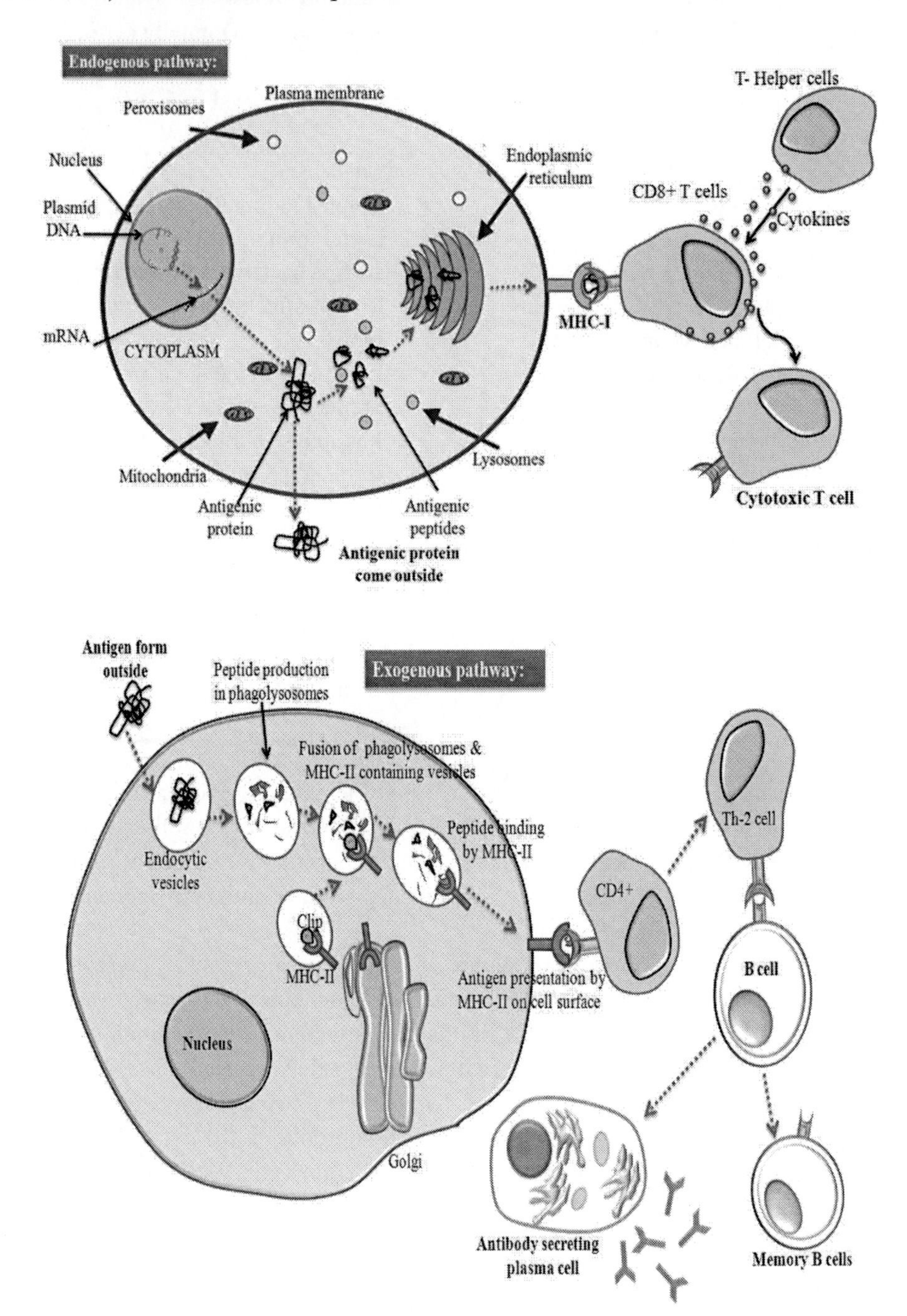


Figure 1. Mechanism of antigen presentation in genetic immunization.

The endogenously synthesized protein antigens however are processed to its peptides fragments in the cytosol by multicatalytic proteasomes. Which are then transported to endoplasmic reticulum by the transporter associated with antigen presentation (TAP). In the endoplasmic reticulum, compatible peptides interact with naïve class I heavy chains and β2-microglobulin-generating transport-competent trimeric complexes which are afterwards delivered to the cell surface where they become accessible to CD8+ T cell. This pathway on antigen presentation is known as cytosolic pathway and occurs for foreign as well as the self proteins. However, the T lymphocytes only recognize the foreign proteins and bind with it.

On the contrary, exogenous proteins are taken up into endosomal compartments from the extracellular milieu and presented on the cell surface in association with MHC class II molecules. In the endosomes antigen is denatured by low pH and degraded by endosomal and lysosomal proteases, revealing peptide segments. These peptide fragments bind to MHC class-II molecules which are trafficked through golgi body vacuole. The resulting class II–peptide complexes are transported to the membrane for detection and help from CD4+ T cells.

During immune recognition the peptide fragments loaded MHC-II are recognized by complementary naïve $CD4^+$ T cells which differentiate into cytokine secreting effector helper cells (Th-1 and Th-2). The Th-2 cell releases cytokines that stimulate pre B cells. These pre B cells on stimulation divide and form antibody secreting plasma cells and long lived memory B cells and remember the antigen for future infection.

On the contrary the peptide fragment loaded onto MHC-I molecule are recognized by naïve CD8+ T cells. These cells stimulate and develop into cytotoxic T cells with the help of cytokines secreted by Th-1 cells. These cytotoxic T cells exhibit cell mediated immunity, binds to the antigen MHC-I complex expressing altered self cell and with the help of effector Th-1 cell destroy the parasite infected cell with the help of perforin. The memory cell in this case can be either CD4+ or CD8+ cells (Seder *et al.*, 2008). In this way the DNA based genetic immunization elicits both humoral and cellular immune responses. Since the DNA-encoded antigen presumably gains access to both the MHC class I and class-II antigen processing pathways.

# 3. Barriers in DNA Vaccine Delivery and Possible Modes to Overcome these Barriers

Though being theoretically most successful there are a number of factors encountered as barriers in the path of its delivery to the target cell (Schatzlein et.al., 2001). The evolutionary barriers are developed to conserve the genetic material of the cells. The exogenous genetic materials are not allowed to enter in to and express with or along master genome of a cell. These biological barriers are classified into two: extracellular and intracellular (Lechardeur and Lukacs, 2002) and presented in the figure 2.

The former one includes: susceptibility to enzymatic degradation by nucleases in the serum and extracellular fluid (Hashida et. al., 1996), vulnerability to inactivation by interactions with blood components (Ogris et. al., 1999) and the possibility of recognition and elimination by immunological defense systems (Patel et. al, 1992). That leads to enzymatic degradation of extracellular DNA by nucleases present in serum and clearing of molecules and its carrier by macrophage of reticuloendothelial system (RES) in liver, spleen and bone marrow. However modification of the DNA by molecular designing (i.e., backbone modification) or entrapment within the carrier system can mask DNA based therapeutics form degradative extracellular environment at the same time the appropriate size of carrier construct can allow entry into fine capillary and tissue (a colloidal size of less than 1μm is ideal) (Dash et. al., 1999). Further, the carrier system can be masked form RES uptake by the attachment of hydrophilic polymer PEG on the surface of carrier (Ogris et. al., 1999).

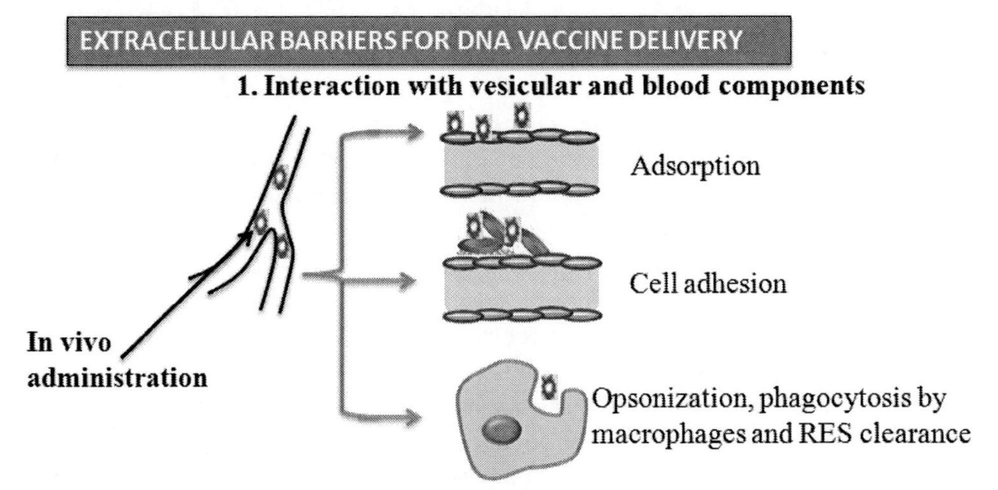


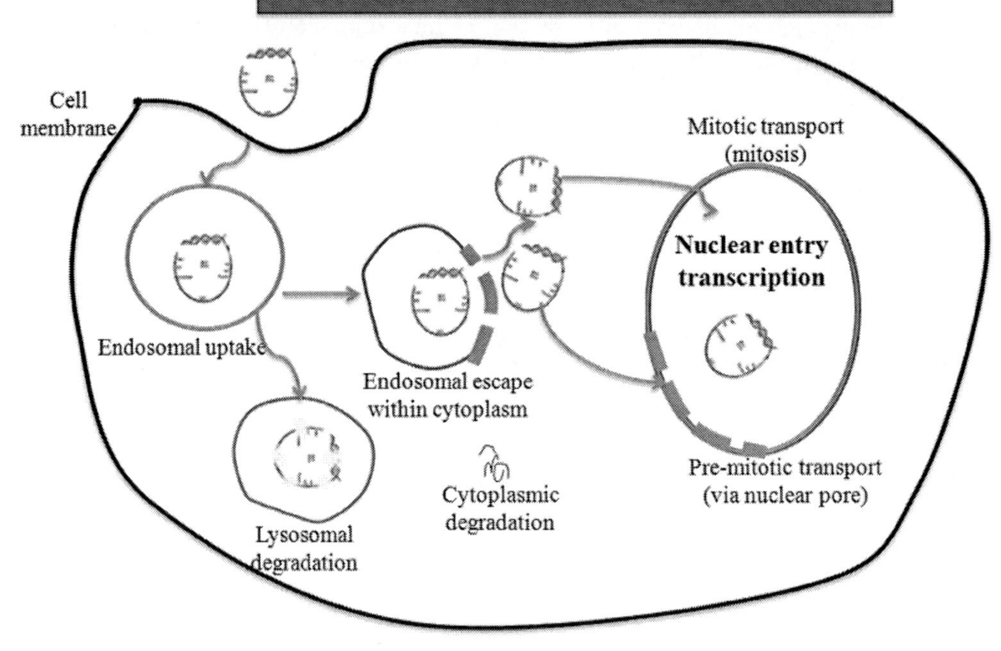


Figure 2. Extracellular and intracellular barriers encountered in the path of DNA vaccine delivery.

The protection of DNA from extracellular barriers does not confirm its expression since the polyanionic hydrophilic nature and large size of DNA based vaccines inhibit its entry through cell membrane. A number of investigations have been made in order to develop a carrier that facilitates the entry of DNA within the cell. The impact of research has establishes the use of cationic polymer having lipophilic nature facilitate entry into the cells. At the same time uptake by non target cell can also be prevented by attachment of some homing device with the DNA carrier complex (eg. galactose, antibodies or transferrin) that will target gene along with its carrier complex to desired therapeutic target (Mislick et. al., 1996).

Even on entry in to the cell that occurs primarily *via* phagocytosis or endocytosis, the degradation of DNA vaccines within the endosomes due to action of endosomal/lysosomal enzymes may lead to failure of effective transfection of genetic construct in to the cytosol. Moreover, endosomal escape and prevention of endosomal digestion can be favoured by the use of pH sensitive lipid or endosomolytic polymers which cause and favour the release of endosomal content into the cytosol. Another hurdle is the degradation of exogenous gene into the cytosol as cytoplasmic nucleases cause its rapid degradation which can be prevented by backbone modification. In addition to above barriers the nucleolar transport of the DNA is basically restricted by the presence of nuclear membranes as the DNA has to enter into the nucleus for transcription to occur. However, in non-dividing cells, the nucleo-cytoplasmic exchange of molecules occurs through the nuclear pore complexes (NPC) that span the nuclear envelope (Ludtke et. al., 1999). Hence, the nuclear envelope acts as a molecular sieve, enabling small aqueous molecules of up to 9 nm in diameter (<17-kDa) to diffuse freely through the NPC (Ohno et. al., 1998). Nevertheless, larger molecules up to 25 nm (>41 kDa) such as plasmid DNA and larger DNA fragments undergo a sequence-specific active transport process involving multiple cellular components (Kreiss et. al., 1999).

# 4. Novel Strategies for DNA Delivery

During last decades, research has been focused on the safe delivery approaches that can bypass the barriers encountered during the delivery of naked DNA and can circumvent the problems associated with the use of viral vectors. Compared to viral systems, nonviral systems are considered to be safe, cheap and able to deliver larger pieces of DNA. Also, these novel carriers avoid DNA degradation and facilitate targeted delivery to APCs to elicit strong humoral and cellular immune responses. Furthermore, the use of these carriers provides a new approach for the induction of secretory responses at mucosal sites and provides an opportunity for immunization via the oral/nasal route. Some of these novel carriers with their uses and route of administrations are summarized in table 3.

## 4.1. Lipid Based Carrier Adjuvants for DNA Vaccination

### *4.1.1. Liposomes*

Liposomes are the most extensively studied versatile carriers for the intracellular delivery of drugs, antigens and DNA (Felnerova *et al.*, 2004). Liposomes are vesicular systems consisting of a hydrophilic core surrounded by concentric lipid bilayers enclosing aqueous components. Both types of drug, hydrophilic as well as hydrophobic, can be entrapped in this system (Gregoriadis, 1978). Liposomes have enormous advantages over a viral delivery system for intracellular delivery of DNA (Patil *et al.*, 2005). For example, (i) they are non-immunogenic due to lack of proteinaceous components (ii) liposomes can be tailored to yield the desired size, surface charge, composition and morphology (iii) they protect DNA from nucleases and improve their biological stability (Rawat *et al.*, 2007).

**Table 3. Commonly used novel carriers for DNA vaccination**

| Formulations | Route of administration | Uses |
|---|---|---|
| *Lipid based carriers* | | |
| DOTMA/cholesterol cationic mannosylated liposomes | IP, IV | Melanoma |
| DOTIM/cholesterol cationic liposomes | IV, IM, ID | Lung tumor |
| Non-ionic surfactant based vesicles (niosomes) | Topical, Oral | HBV |
| *Polymeric microparticles* | | |
| Poly(lactide-co-glycolide) (PLGA) | IM | HIV, Solid tumors |
| PLGA w/cetyltrimethylammonium bromide (CTAB) | | Measles virus, HCV |
| PLGA w/PEI coating | ID, IM | B-cell lymphoma |
| PLGA w/PBAE | ID | Tumor antigen |
| *Polymeric nanoparticles* | | |
| Poly(ethylenimine)–mannose (PEI–man) (DermaVir patch) | Transdermal | HIV |
| Poly-L-lysine (PLL) | ID | HIV |
| PLL-coated polystyrene | ID | Model tumor antigen |
| Chitosan | IN, oral, pulm. | Allergy, RSV, tuberculosis |

Abbreviations: IM: intramuscular, ID: intradermal, IN: intranasal, pulm: pulmonary, IV: intravascular, IP: intraperitoneal .

Cationic liposomes, the most promising carrier systems for gene therapy, have been used for the delivery of various plasmids, oligonucleotides, DNA and RNA to a variety of cells (Nakanishi and Noguchi 2001; Reddy *et al.* 2002; Kamiya *et al.* 2002; Chiu *et al.* 2006). Cationic liposomes were first reported by Felgner *et al.* in 1987 for the efficient transfection of eukaryotic cells. These liposomes were composed of cationic lipid [N-[1-(2,3-dioleyloxy) propyl]-N,N,N,-trimethylammoniumchloride, DOTMA] and zwitterionic lipid (DOPE) in 1 : 1 ratio. Other cationic lipids commonly used in cationic liposomes are DOTAP, DOTMA, DOTIM, DC-CHOL, DDAB etc. The chemical structures of some of these are shown in figure 3. Commonly used zwitterionic lipids are DOPE and cholesterol. DOPE shows fusion activity with endosomal/lysosomal membranes and help in the endosomal escape of the liposomal contents. It also reduces the cytotoxicity of cationic lipids (Gustafsson *et al.* 1995; Lechardeur *et al.* 2005).

Cationic lipids are debatably the most common nonviral transfection reagent. These impart a positive charge to the liposomes that helps in the complexation and condensation of DNA and also in the interaction with the cell surface (Ahmed *et al.*, 2005). Previously there were three models proposed for the interaction of a cationic liposome with the cell and release of DNA and oligonucleotide into the cytosol (Wrobel and Collins, 1995), (i) liposome-cell fusion within or destabilization of the endosomes (ii) direct fusion with plasma membrane (iii) transfer of the lipid-DNA complex across the cellular membrane into the cytosol.

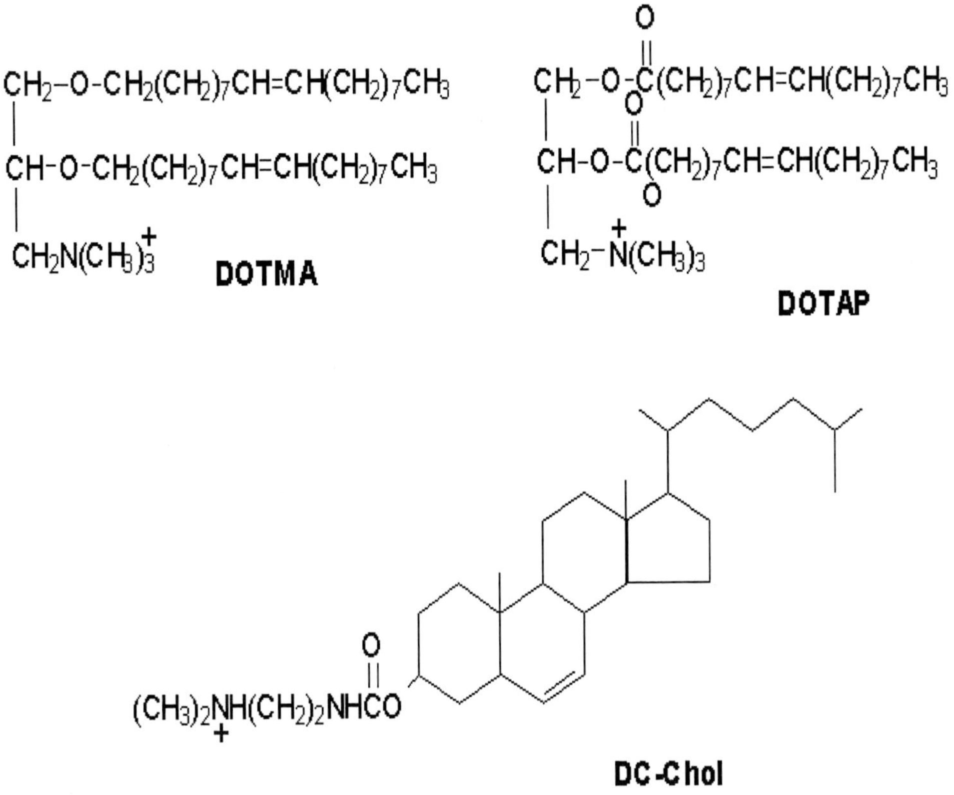

$CH_2$-O-$CH_2(CH_2)_7CH=CH(CH_2)_7CH_3$
CH-O-$CH_2(CH_2)_7CH=CH(CH_2)_7CH_3$
$CH_2N^+(CH_3)_3$
DOTMA

$CH_2$-O-C(O)-$(CH_2)_7CH=CH(CH_2)_7CH_3$
CH-O-C(O)-$(CH_2)_7CH=CH(CH_2)_7CH_3$
$CH_2$-$N^+(CH_3)_3$
DOTAP

$(CH_3)_2N^+H(CH_2)_2NHC(O)$
DC-Chol

Figure 3. Structures of cationic lipids commonly used in cationic liposomes.

Later Xu and Szoka, (1996) gave a hypothetical model to explain the mechanism of release of cationic lipid/DNA complexes from endosomes. In this model, cationic lipid/DNA complex after internalization first destabilizes the endosome membrane. After destabilization, the negatively charged lipids in the cytosolic phase move to the endosomal phase via a flip-flop mechanism. The anionic lipids then diffuse via lateral diffusion to form neutral ion pairs with cationic lipids. As a result, DNA, which was bound to the cationic lipids electrostatically, is displaced and released into the cytosol.

Study showed that cationic liposome-mediated DNA immunization induced stronger hepatitis C virus (HCV) non-structural protein 3 (NS3) -specific immune responses than immunization with naked DNA alone. Cationic liposomes, composed of DDAB and equimolar neutral lipid, egg yolk phosphatidylcholine (EPC), induced the strongest antigen-specific Th1 type immune responses among the cationic liposome investigated, whereas the liposomes composed of 2 cationic lipids, DDAB and DOEPC, induced an antigen-specific

Th2 type immune response. All cationic liposomes used in this study triggered high-level, non-specific IL-12 production in mice, a feature important for the development of maximum Th1 immune responses (Jiao *et al.*, 2003).

Plasmid DNA can also be encapsulated within a lipid vesicle called a dehydrated-rehydrated vesicle (DRV). Dehydration rehydration vesicle (DRV) technology has also been used for the delivery of plasmid DNA for the purposes of immunisation (Gregoriadis *et al.*, 1999). DRVs are produced through the process of freeze-drying of lipoplexes (which is thought to increase the association of plasmid DNA with the flattened liposomal vesicles). Subsequent rehydration leads to the formation of a DRV with entrapped plasmid. DRV formulation is able to generate submicron-sized liposomes incorporating most of the DNA in a way that prevents DNA displacement through anion competition, indicating that much of the DNA is entrapped within the aqueous compartments in between bilayers (Perrie *et al.*, 2000).

DRVs, compared to naked DNA and cationic lipoplexes, have shown the ability to generate improved cellular immunity and secretion of greater IgG1 levels in mice (Gregoriadis *et al.*, 1997). Various lipids have been shown to increase vaccine potency when added to these formulations and have also been shown to enhance oral delivery of DRV formulations (Perrie *et al.*, 2001, 2002). Gregoriadis *et al.*, (1997), for the first time showed that intramuscular immunization of mice with pRc/CMV-HB(s) (encoding the S region of hepatitis B antigen; HBsAg) entrapped into positively charged (cationic) liposomes leads to greatly improved humoral and cell-mediated immunity. These cationic liposome-entrapped DNA vaccines generate titres of anti-HBsAg IgG1 antibody isotype in excess of 100-fold higher and increased levels of both IFN-gamma and IL-4 when compared with naked DNA or DNA complexed with preformed similar (cationic) liposomes. It is likely that immunization with liposome-entrapped plasmid DNA involves antigen-presenting cells locally or in the regional draining lymph nodes. Animal experiments have shown that immunization through intramuscular or the subcutaneous route with liposome-entrapped plasmid DNA encoding the hepatitis B surface antigen leads to much greater humoral (IgG subclasses) and cell mediated (splenic IFN-$\gamma$) immune responses than with naked DNA (Perrie *et al.*, 2004). In other experiments with a plasmid DNA encoding a model antigen (ovalbumin), a CTL response was also observed. These results could be explained by the ability of liposomes to protect their DNA content from local nucleases and direct it to APCs in the lymph nodes draining the injected site (Gregoriadis *et al.*, 2002).

It has been shown that the polymer modified liposomes offer potential for oral administration of plasmid DNA and able to elicit markedly enhanced transgene specific cytokine production following *in vitro* restimulation of splenocytes with recombinant antigen (Somavarapu *et al*, 2003). Also, coating of polymer on to the surface of liposome/DNA complexes containing pRc/CMV-HBs(S) may potentially enhance *in vivo* delivery of plasmid DNA to antigen presenting cells and thereby facilitate enhanced immune responses against encoded protein. Modification of lipid/DNA complexes by the polymer poly(d,l-lactic acid) was found to be consistently and significantly more effective than either unmodified liposomal DNA or naked DNA in eliciting transgene-specific immune responses to plasmid-encoded antigen when administered by the subcutaneous route. In addition, the polymer-modified formulations delivered through this route were more effective than naked DNA delivered by the intramuscular route in inducing antibody responses (Bramwell *et al.*, 2002).

Intranasal immunization with liposome encapsulated influenza hemagglutinin (HA) DNA (pcI-HA10) vaccine induced T cell proliferation, indicative of CD4+ activity, in addition to increasing serum IgG and IgA titres, suggestive of humoral immune responses. Mice challenged with a lethal dose of influenza virus following intranasal immunization with liposome encapsulated vaccine were completely protected (Wang *et al.*, 2004). Recently, our group developed polymer modified liposomes for the delivery of plasmid DNA encoding Hepatitis B surface antigen (Khatri *et al.*, 2008a). Plasmid pRc/CMV-HBs(S) encapsulated liposomes were prepared by dehydration-rehydration method and subsequently coated with glycol chitosan by simple incubation method. Nasal administration of glycol chitosan modified liposomes resulted in serum anti-HBsAg titre which was less compared to that elicited by naked DNA and alum adsorbed HBsAg administered intramuscularly, but the mice were seroprotective within 2 weeks and the immunoglobulin level was above the clinically required protective level (>10 mIU/ml) suggesting successful generation of systemic immunity (Figure 4A). In addition, only glycol chitosan modified liposomes treated mice had high levels of secretory IgA in nasal, salivary and vaginal secretions indicating successful induction of mucosal immunity as well (Figure 4B). The significant levels of both IL-2 and IFN-γ (Figure 4C and 4D) were measured in mice immunized with pDNA administered IM and with pDNA loaded glycol chitosan modified liposomes as compared to those recorded for alum adsorbed HBsAg and pDNA administered intranasally. Both Th1 dependent cytokines and their high levels are evidenced for the strong cell-mediated immune response elicited by liposomal formulations administered intranasally.

The immunological efficacy, in case of genetic immunization, can be improved by delivering antigen gene directly to the cytoplasm via membrane fusion using fusogenic liposomes. Accordingly, immunization with fusogenic liposomes containing the OVA-gene induced potent OVA-specific Th1 and Th2 cytokines production. Additionally, OVA-specific CTL responses and antibody production were also recorded in systemic compartments including the spleen, upon immunization with the OVA-gene encapsulating fusogenic liposomes (Yoshikawa *et al.*, 2004). Mannosylated cationic liposomes were developed for targeted delivery of plasmid DNA (pDNA) to antigen-presenting cells (APCs), and the results verified that Man lipoplex induces significantly higher pUb-M gene transfection into dendritic cells and macrophages than unmodified lipoplex and naked DNA and it also strongly induces CTL activity against melanoma and prolongs the survival after tumor challenge compared with unmodified liposomes (Lu *et al.*, 2007). Also in a different study, following IV administration, OVA mRNA expression and MHC class I-restricted antigen presentation on CD11c+ cells and inflammatory cytokines, such as TNF-α, IL-12, and IFN-γ, that can enhance the Th1 response of the Man liposome/pCMV-OVA complex were higher than that of naked pCMV-OVA and that complexed with DC-Chol liposomes.

Also, the spleen cells from mice immunized by IV administration of the Man liposome/ pCMV-OVA complex showed the highest proliferation response and IFN-γ secretion (Hattori *et al.*, 2004).

In other study, cationic transfersomes composed of cationic lipid DOTMA and sodium deoxycholate as constitutive lipids were prepared and plasmid DNA encoding HBsAg was incorporated in the cationic transfersomes using charge neutralization method. Results revealed that DNA loaded cationic transfersomes elicited significantly higher anti-HBsAg antibody titer and cytokines level as compared to naked DNA. The study signifies the potential of cationic transfersomes as DNA vaccine carriers for effective topical immunization (Mahor *et al.*, 2007).

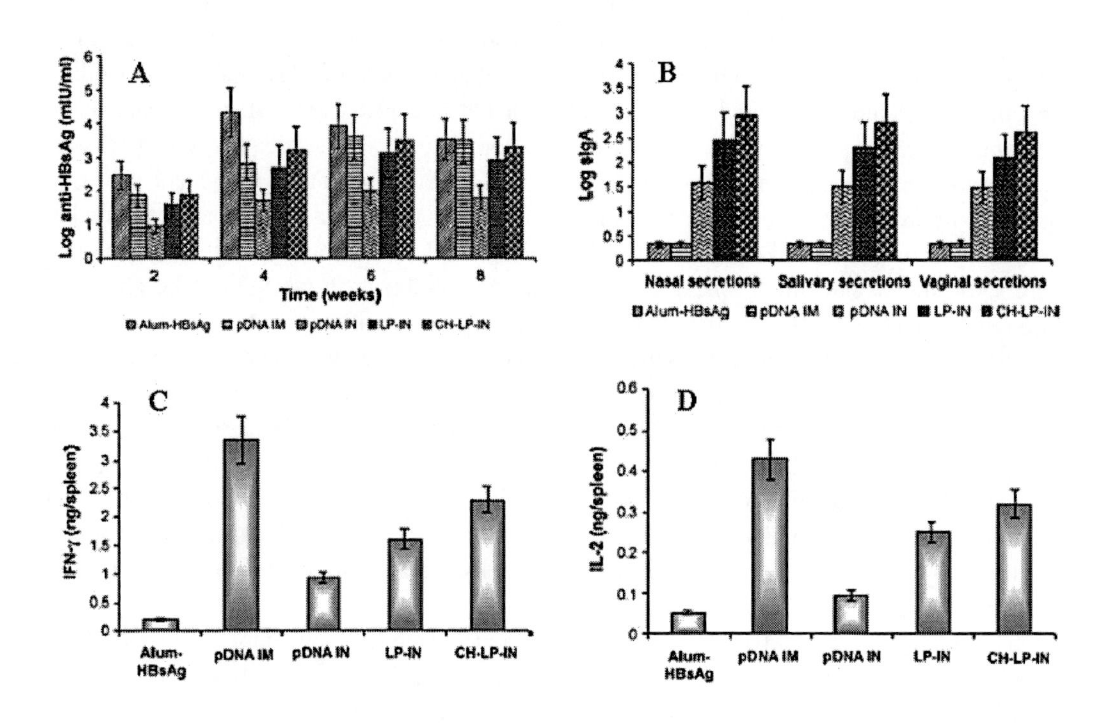


Figure 4. Serum anti-HBsAg profile (A), secretory IgA levels in nasal, salivary and vaginal secretions (B), IFN-γ (C) and IL-2 (D) levels in spleen, of mice immunized with various formulations. From Khatri *et al.*, 2008a.

### *4.1.2. Virosomes*

To obtain improved efficiency of liposomal interaction with the cellular target for the transportation of the molecules directly into cells a novel version of liposomes, virosomes have been developed. By and large 'Virosomes' are liposomes epiked with viral proteins extracted from virion envelopes, which acquire viral functions like cell surface attachment as well as fusogenicity to cellular and organelle membranes but are free from infertility. Virosomes can be considered as hybrids between viral and liposomal delivery systems, combining the characteristics of cellular interactions of viral vectors with the safety of liposomal delivery systems. These are reconstituted membrane vesicles prepared from enveloped viruses, but lack the genetic material of the native virus. Ultrastructurally, virosomes are spherical unilamellar vesicle with a mean diameter of 150 nm and short surface projections of 10-15 nm (Tyagi *et al.*, 2008).

The newer version of liposomes, virosomes seems to possess many ideal properties for *in vivo* gene transfer such as no limitation of encapsulated DNA size, high efficiency for cytosolic delivery, non immunogenicity, simplicity in handling and brevity of incubation time. Okamoto *et al.* (1997) demonstrated increased antibody response when using

Hemaglutinating Virus Japan (HVJ) fusogenic peptides to deliver plasmid DNA i.m. in a mouse model while naked DNA was ineffective.

Kaneda *et al.*, (1999) constructed hybrid-type of liposomes with a fusogenic envelope derived from hemagglutinating virus of Japan (HVJ; Sendai virus). In the preparation HVJ virosomes, liposomes acquire viral function by their fusion with UV-inactivated HVJ. The procedure is summarized in figure 5.

The HVJ liposomes can encapsulate DNA smaller than 100 kb, RNA, oligodeoxy-nucleotides including antisense. The lipid membrane also provides an additional protection of the encapsulated genetic molecules from the enzymatic degradation in serum as well as body fluid (Hirano *et al.*, 1998). Immunogenicity and cytotoxicity associated with HVJ-liposomes was also reported. HVJ-liposomes shown to allow the repeated injections as anti-HVJ antibody generated in the rat was not sufficient to neutralize HVJ-liposomes, in addition, cytotoxic T cells recognizing HVJ were not detected in the rat transfected repeatedly with HVJ-liposomes. There was no significant cell damage and dysfunction of target organ detected when $10^{10}$ to $10^{11}$ particle of HVJ liposomes were injected into portal veins of rats. In a broad sense, virosomes retain both the infectivity potential of viral vectors as well as general features of non-viral vectors (Dzau *et al.*, 1996). Virosomes are versatile antigen carriers and can be engineered to perform various tasks in cancer immunotherapy. Preclinical data have fostered the development of innovative clinical protocols.

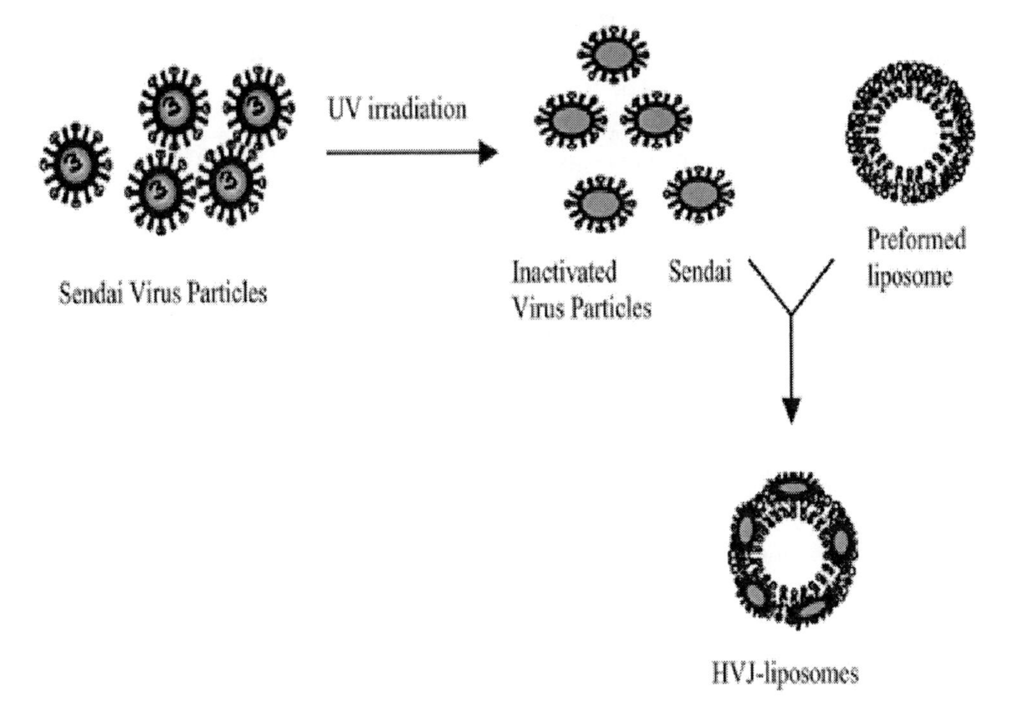


Figure 5. Diagram showing formation of HVJ-Liposomes.

In the context of DNA delivery, modified immunopotentiating reconstituted influenza virosomes (IRIVs) composed of spherical, unilamellar vesicles, and prepared using a mixture of natural and synthetic phospholipids and 10% envelope phospholipids originating from influenza A/Singapore/6/86 and influenza surface glycoproteins, have been used for the delivery of DNA (encoding mumps virus haemagglutinin) to dendritic cells (Cusi *et al.*, 2004) showing uptake of fluorescently labelled DNA by cells with dendritic phenotype *in vivo*. Subsequent work from the same group showed an enhancement of immune responses against carcinoembryonic antigen using virosomes which delivered a gene encoding the costimulatory molecule CD40L (Cusi *et al.*, 2005). Hence, immunopotentiating reconstituted influenza virosomes will be assessed in breast and melanoma immunotherapy, and may contribute to the development of clinically effective cancer vaccines and ultimately improve patient outcomes (Adamina *et al.*, 2006). Cusi *et al.* 2000 demonstrated a new delivery system based on the modified immunopotentiating reconstituted influenza virus (IRIV). The influenza DOTAP cationic virosomes used in the study were exogenously loaded with plasmid DNA vector expressing the mumps virus hemagglutinin or the fusion protein. The administration of this DNA vaccine in combination with the mucosal adjuvant Escheriagen via the intranasal route was efficient for inducing an immune response, both mucosally and systemically, in mice. The production of IgG2a mumps virus-specific antibodies and the secretion of IL-10 by antigen-specific T cells further indicated that not only Th1 but also Th2 responses were induced by this DNA vaccine formulation (Cusi *et al.*, 2000). In a different study, it has been described that DNA expressing the parathyroid hormone related peptide (PTH-rP), a protein secreted by prostate and lung carcinoma cells can be coupled to virosomes or IRIVs. Mice immunised and boosted twice intranasally with these DNA virosomes developed a low PTH-rP specific CTL response and this response could be slightly increased by subcutaneous administration of interleukin-2 for 5 days per week during the entire immunisation/booster protocol (6 weeks). In HLAA02.01 transgenic mice, similar responses were induced suggesting that the vaccine can be employed for immunotherapy of human cancers and metastases overexpressing PTH-rP (Correale *et al.*, 2001; Scardino *et al.*, 2003).

A novel tuberculosis (TB) vaccine comprising of a combination of the DNA vaccines expressing mycobacterial heat shock protein 65 (HSP65) and interleukin 12 (IL-12) was delivered using the hemagglutinating virus of Japan (HVJ)-liposome (HSP65 + IL-12/HVJ). This vaccine provided remarkable protective efficacy in mouse and guinea pig models compared to the conventional BCG vaccine, on account of an induction of the CTL activity and improvement of the histopathological tuberculosis lesions, respectively. Furthermore, the studies were extended on cynomolgus monkey model, which is currently considered to be the best animal model of human tuberculosis. This novel vaccine provided a higher level of the protective efficacy than BCG in regards to assessment of mortality, the ESR, body weight, chest X-ray findings and immune responses. Furthermore, the combination of HSP65 + IL-12/HVJ and BCG by the priming-booster method showed a synergistic effect in the TB-infected cynomolgus monkey resulting into 100% survival (Okada *et al.*, 2007). Following a single gene gun vaccination with the combination of Hsp65 DNA and mIL-12 DNA provided a remarkably high degree of protection against challenge with virulent *Mycobacterium tuberculosis*; bacterial numbers were 100-fold lower in the lungs compared to BCG-vaccinated mice. The HVJ-liposome improved the protective efficacy of the Hsp65 DNA vaccine compared to gene gun vaccination. This protective efficacy was associated with the emergence of IFN-γ-secreting T cells and activation of proliferative T cells and cytokines (IFN-γ and IL-2) production upon stimulation with Hsp65 and antigens from *M. tuberculosis* (Yoshida *et al.*, 2006).

## 4.2. Polymers and Polymeric Carriers Systems

Another method which has been widely studied for intracellular delivery of DNA is based on cationic polymers. Polymers, bearing groups that are protonated at physiological pH, have been used as gene carriers, forming complexes with DNA called polyplexes due to the interaction of cationic charge of the polymer and the negative charge of DNA. These complexes protect DNA from enzymatic degradation. Due to the cationic charge of the polyplex it adheres to the cell membrane and subsequently taken up by the cells via endocytosis (Wiethoff and Middaugh 2003). Commonly used cationic polymers include poly-L-lysine (PLL), polyethyleneimine (PEI), chitosan, poly (2-dimethylaminoethyl)-methoxy (pDMAEMA), polybrene, tetraminofullerene, cationic polysaccharides and cationic dendrimers (Azzam *et al.* 2004). Structures of some of these polymers are presented in figure 6. The high efficiency of cationic polymer is due to buffering effect or 'proton sponge effect' of the polymers by the presence of amino groups in the molecules (Patil *et al.*, 2005). It was suggested that when PEI-based polyplexes reach the endosome, protonation of amine groups occurs in the acidic environment of the vesicles. By protonation, polymer swellings take place, and at the same time the endosomes also swells due to osmotic imbalance. The combined swelling ruptures the endosomal membrane and releases the content into the cytosol (Cho *et al.* 2003). This proton sponge hypothesis is valid not only for PEI but also for other polymers containing amine groups with pKa at or below physiological pH, i.e. pDMAEMA (Van de Wetering *et al.* 1999), polyamidoamine (PAMAM) dendrimers (Haensler and Szoka 1993), histidylated polylysine (Midoux *et al.* 1998) and lipopolyamine (Ahmed *et al.* 2005).

### *4.2.1. Polyethylenimine*

One of the most potent polymers for gene delivery at present is poly(ethyleneimine) (PEI). PEIs were first introduced by Behr in 1995 (Boussif *et al.*, 1995), and have become one of the most efficient nonviral gene delivery systems. PEI is a group of synthetic polymeric polycation carriers known to be efficient in the transport of macromolecules like oligonucleotides and plasmid DNA in a variety of cells and animal models. PEI with a molecular weight of 25 KDa (25K pEI) displays a high transfection efficiency, probably due to efficient endosomal escape. However, a considerable toxicity has been reported, whereas low molecular weight (LMW) pEI is less toxic but shows almost no transfection (Godwey *et al.*, 1999, Fischer *et al.*, 1999). Therefore, several investigators synthesized (highly) branched polymers consisting of LMW PEI and degradable cross-links. Linear and branched PEI can induce the condensation of DNA to nanoparticles. The linear and branched nature of PEI as well as its molecular weight plays an important role in DNA condensation and transfection efficiency. The buffering capacity of PEI is believed to contribute to its ability to deliver DNA within cells without degradation. Unlike other polymers, PEI possesses intrinsic endosomolytic activity and does not need any endosomolytic agent to support escape from endosome. Poor solubility of the DNA/PEI complex at physiological pH and toxicity of PEI in animal model studies are the major constraints in using this polymer as a gene carrier (Brown *et al.*, 2001). Free PEI may harm cells, but when bound to DNA the detrimental effects are greatly decreased.

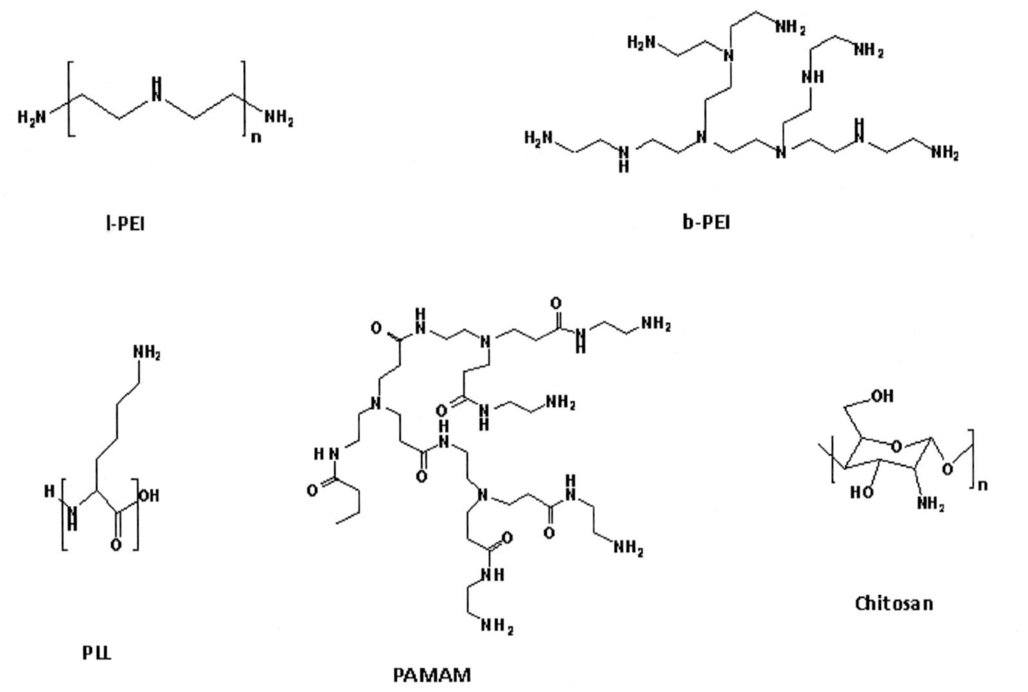

l-PEI

b-PEI

PLL

PAMAM

Chitosan

Figure 6. Structures of most commonly used cationic polymers.

A possible reason for toxic effects of PEI on cells is permeabilizing effect of PEI on membranes. However, data from various studies suggested that low concentrations of PEI could not harm plasma membranes (Oku *et al.*, 1986; Lambert *et al.*, 1996; Klemm *et al.*, 1998).

In order to investigate the effects of the protonation properties of the polymer, PEI derivatives were synthsized by acetylating PEI with increasing amounts of acetic anhydride to yield polymers with 15%, 27%, and 43% of the primary amines modified with acetyl groups. Acetylation of PEI decreased the "physiological buffering capacity," and greatly enhanced the gene delivery activity of the polymer. The mechanism is not yet understood, but the enhancement may be caused by more effective polyplex unpackaging, altered endocytic trafficking, and/or increased lipophilicity of acetylated PEI-DNA complexes (Forrest *et al.*, 2004). Further, increasing the degree of acetylation, gene delivery activity continued to increase (up to 58-fold in HEK293) with acetylation of up to 57% of primary amines but decreased at yet higher degrees of acetylation.

PEI-based formulations have proven stable during nebulization and result in transfection of a very large proportion of epithelial cells throughout the airways (though the level of transgene expression per cell may be relatively low), though lower levels of transfection, throughout the lung parenchyma. Most importantly, therapeutic responses have been obtained in several animal lung tumor models when PEI-based complexes of p53 and IL-12 genes were delivered by using aerosol. This approach may be useful as a means of localized genetic immunization (Densmore *et al.*, 2000). The administration of PEI-DNA complex expressing HIV-glycoprotein 120 (gp120) antigen (PEI-pgp120) resulted in rapid elevation of serum levels of IL-12 and IFN-gamma. Furthermore, a single administration of PEI-pgp120 complex elicits a number of gp120-specific CD8+ T cells 20 times higher than that elicited by three intramuscular injections of naked DNA. Interestingly, it has been found that systemic vaccination with PEI-pgp120 induced the protective immune responses against both systemic and mucosal challenges with a recombinant vaccinia virus expressing a gp120 antigen (Garzon *et al.*, 2005).

Particle-bound plasmid DNA may have utility in genetic immunization by intravascular delivery to the lung and potentially to other organs and tissues. PEI can be conjugated to serum albumin and the conjugate was aggregated by heating to produce particles of 25-100 μm. The resulting particles bound plasmid DNA when injected IV in mice, the particles distributed in the peripheral lung tissue of the alveolar interstitium. Particle-bound luciferase plasmid transfected a variety of cell lines *in vitro*, and after IV injection, gene expression was detected exclusively in the lung. Using human growth hormone as the encoded foreign Ag for immunization, IV injection of the particle-bound plasmid elicited both pulmonary mucosal and systemic immune responses, whereas naked DNA injected either IV or IM elicited only systemic responses (Orson *et al.*, 2000).

Recently, it has been demonstrated that PEI can also be used as an adjuvant for nasal administration for the induction of potent systemic and mucosal responses, both cellular as well as humoral (Torrieri-Dramard *et al.*, 2010). Result of the study showed that PEI can improve the efficiency of gene transfer 1,000-fold in the respiratory track following intranasal administration of luciferase-coding DNA. Using PEI formulation, intranasal vaccination with DNA-encoding hemagglutinin (HA) from influenza A H5N1 or (H1N1) 2009 viruses induced high levels of HA-specific immunoglobulin A (IgA) antibodies that were detected in bronchoalveolar lavages (BALs) and the serum. No mucosal responses could be detected after parenteral or intranasal immunization with naked-DNA. Furthermore, intranasal DNA vaccination with HA from a given H5N1 virus elicited full protection against the parental strain and partial cross-protection against a distinct highly pathogenic H5N1 strain that could be improved by adding neuraminidase (NA) DNA plasmids.

In different studies, it has been shown that the pulmonary delivery of plasmid DNA formulated with polyethyleneimine (PEI-DNA) induced robust systemic CD8+ T-cell responses that were comparable to those generated by IM immunization (Bivas-Benita *et al.*, 2010). Interestingly, it was observed that the pulmonary delivery of PEI-DNA elicited a 10-fold-greater antigen-specific CD8+T-cell response in lungs and draining lymph nodes of mice than that of IM immunization. The functional evaluation of pulmonary CD8+ T cells revealed that they produced type I cytokines, and pulmonary immunization with PEI-DNA induced lung-associated antigen-specific CD4+ T cells that produced higher levels of IL-2 than those induced by IM immunization. Pulmonary PEI-DNA immunization also induced CD8+ T-cell responses in the gut and vaginal mucosa. Results suggested that pulmonary PEI-DNA immunization might be a useful approach for immunizing against pulmonary pathogens and might also protect against infections initiated at other mucosal sites.

#### *4.2.2. Polylysine*

Poly(L-lysine) (PLL) is one of the polymers that have been thoroughly investigated as a non-viral gene delivery vector (Wagner *et al.*, 1998). However, complexes of PLL and DNA have relatively low transfection activity and a rather high toxicity, especially when high molecular weight PLL (Mw 25 kDa) is used. PLL with different molecular weights has been studied in physicochemical and biological experiments for gene delivery (Wolfert *et al.*, 1999). Although PLL is biodegradable, still it exhibits modest to high toxicity. The polyplexes of PLL are efficiently taken up into cells; however transfection efficiencies remain considerably lower. A potential reason for this is the lack of amino groups with a pKa between 5 and 7, thus allowing no endosomolysis and hence low levels of transgene expression. The inclusion of targeting moieties or co-application of endosomolytic agents like fusogenic peptides (Wagner *et al.*, 1992) or chloroquine (Pouton *et al.*, 1998) may improve reporter gene expression. In addition, attachment of histidine or other imidazole containing structures to PLL (i.c., pKa around 6, thus possessing a buffering capacity in the endolysosomal pH range) showed a significant enhancement of reporter gene expression compared to unmodified PLL (Benns *et al.*, 2000; Midoux and Monsigny, 1999; Faiac *et al.*, 2000).

In another approach, McKenzie *et al.* (2000) used the reductive intracellular environment as a tool to degrade polyplexes based on lysine residues. For that purpose, lysine oligomers with or without histidine residues were terminated on both ends by cysteine residues. After complexation with DNA, these oligomers spontaneously linked by covalent disulfide bonds, thus stabilizing the complexes and enhancing gene expression. The histidine residues were shown to provide buffering capacity to further enhance endosomal escape and *in vitro* gene expression. It is assumed that the disulfide bonds are reduced by high intracellular concentrations of glutathione, thus destabilizing the complexes and releasing the DNA.

A polymeric condensing system based on PLL has been tested in a Phase 1–2 clinical trials to deliver cystic fibrosis transmembrane regulator CFTR plasmid by the intranasal route (Konstan *et al.*, 2004). The complex consists of a single plasmid compacted with polyethylene glycol (PEG)-substituted 30-mer lysine polymer (CK30). The plasmid (8.3 kbp) encodes the CFTR protein. The plasmid is condensed into rod-like nanoparticles with a diameter of 12-15 nm and a length of 100-150 nm (Ziady *et al*, 2003). The DNA nanoparticles thus formed are unique for a non-viral delivery in that they can be highly concentrated without aggregation in saline. They are stable in serum and can be combined with liposomes and used to transfect post-mitotic cells. Once inside cells, they can directly cross into the nucleus via the 25 nm nuclear pore complex (Liu *et al.*, 2003). In the Phase 1-2 clinical trial, administration of compacted DNA nanoparticles to the nasal epithelium of CF patients was safe. There were no significant adverse events and cell transfection was detected in the nasal epithelial cells.

Shimizu *et al.*, (1996) described an improved method to produce a conjugate of anti-erythrocyte growth factor (EGF) receptor monoclonal antibody with polylysine via thio-ether bonds. The resulting antibody/polylysine conjugate was found to be a much more stable DNA carrier than the previous conjugate formed via disulfide bonds. The conjugate has been designated as an "immunoporter" and the immunoporter/DNA complex as an "immunogene." The fluorescent microscopic observation showed that the immunoporter as well as immunogene bound specifically to the EGF receptors on the cell surface, and the loaded reporter gene, such as beta-galactosidase (beta-GAL), was detected in the cell nucleus at 2 hours after transfection. The enzyme activity from the beta-GAL gene was detected at 12 hours and increased for 3 to 5 days. Thus, the immunogene approach was successful in delivering therapeutic genes to EGF receptor over expressing tumor cells. Further, introduction of lysine (Lys) unit as a DNA anchoring moiety into the amino acid sequence in poly(ethylene glycol)-b-cationic poly(N-substituted asparagine) with a flanking N-(2-aminoethyl)-2-aminoethyl group (PEG-b-Asp(DET)) resulted in PEG-b-P[Lys/Asp(DET)], in which the Asp(DET) unit acts as a buffering moiety inducing endosomal escape with minimal cytotoxicity. The introduction of Lys units into the catiomer sequence facilitated cellular uptake and a 100-fold higher level of gene expression with PEG-b-P [Lys/Asp(DET)]/DNA polyplex micelles prepared even at a lowered N/P (Miyata *et al.*, 2007).

It is well established that transfection of DCs can lead to an increase in antitumoral immunostimulatory capacity of DCs and may have a major impact on immunotherapeutic protocols for patients with cancer. DCs were transfected with the CIITA gene using a novel transfection technique. The vector system consisted of a plasmid bound to an adenovirus via PLL, which is covalently bound to a UV-irradiated adenovirus. After transfection, expression of MHC class II on DCs increased from 27% to 75% on day 2. Further, cytotoxicity of effector cells against tumor cells increased after co-culture with transfected DCs to 63% compared to 15% with effector cells co-cultured with irrelevantly transfected DCs (Marten *et al.*, 2001).

Recently, it has also been investigated that PLL-coated polystyrene nanoparticles complexed to plasmid DNA (encoding for OVA) elicited strong humoral and cellular response superior to immunization with naked DNA (Minigo *et al.*, 2007,). Moreover, inhibition of tumour growth was observed in mice immunised with the nanoparticle DNA vaccine following challenge with an OVA-expressing tumour cell line (EG7.OVA). In this study, PLL was as a cationic linker for electrostatic interaction with negatively charged DNA. Further, no immune responses were observed after two immunizations with pDNA/nanoparticles mixture without PLL, indicating that the PLL linker is essential for the efficacy of the nanoparticle vaccine.

#### *4.2.3. Chitosan*

Chitosan is a good candidate for gene delivery system because of positive charge it can complex negatively charged DNA and can protect DNA from nuclease degradation. It has advantages of not necessitating sonication and organic solvents for its preparation, therefore minimizing possible damage to DNA during complexation. DNA-loaded chitosan microparticles were found to be stable during storage. Chitosan can be obtained in a range of molecular weights from oligomeric materials containing a few units of glucosamine through to higher molecular weight materials of more than 200,000 Daltons. In pharmaceutical applications, the higher molecular weights (50,000-500,000 Daltons) are normally preferred. Chitosan can also be obtained in different degrees of deacetylation, but the materials that have a deacetylation between 60 and 90% are normally preferred. Because chitosan carries a positive charge, it can be used to interact with negatively charged surfaces as well as with other negatively charged materials including pharmaceuticals (Khatri *et al.*, 2008c).

Nanoparticles produced with chitosan of Mw 213 kDa and degree of acetylation 88% showed the highest zeta potential (+23 mV), cellular uptake (4.1 Ag/mg protein) and transfection efficiency (12.1%), while chitosan vector with Mw of 213 kDa and degree of acetylation 46% showed the lowest cellular uptake (0.4 Ag/mg protein) and transfection efficiency (0.05%). Confocal microscopic study further suggested that the chitosan-complexed DNA successfully escaped from the endo-lysosomal compartment for nuclear translocation and expression (Huang *et al.*, 2005).

Chitosan in particular has been investigated as a nanoparticulate carrier for DNA vaccines due to its mucoadhesive and drug delivery properties across mucosal surfaces. As the structurally related chitin is recognized by MPs in the induction of allergy, it is possible that chitosan triggers innate immune pattern recognition receptors on APCs (Reese *et al.*,

2007). Chitosan–DNA nanoparticles induce stronger activation of DCs than naked DNA alone (Bivas-Benita *et al.*, 2004). Chitosan may be potentially useful for DNA vaccines to induce immune tolerance by shifting the TH1 versus TH2 response, and oral immunization with chitosan nanoparticles modulates anaphylactic allergic immune responses (Roy *et al.*, 1999). Chitosan derivatives have been shown to increase epithelial permeability, perhaps by affecting tight junctions (Thanou *et al.*, 2000). Oral delivery of chitosan–DNA nanoparticles (150 nm) in size induced gene expression in the intestinal epithelium. Intranasal (IN) administration of similar chitosan–DNA nanoparticles protected against challenge with respiratory syncytial virus (RSV) due to generation of efficient mucosal antibody responses and cell-mediated responses (Kumar *et al.*, 2002). Larger (~300nm) chitosan–DNA particles were also shown to induce both humoral and cellular immune responses that are protective against RSV challenge following IN and ID administration (Iqbal *et al.*, 2003).

Chitosan particles have tremendous potential as vectors for the transfer of DNA into mammalian cells. Cellular transfection by the chitosan-pGL3-control particles showed a sustained expression of the luciferase gene for about 10 days. Agarose gel electrophoresis and displacement experiments using polyaspartic acid indicated a probable multiple interaction between DNA and chitosan. No toxic effect on the mammalian cells was found with chitosan (Li *et al.*, 2003). Moreover, lyophilized chitosan DNA nanoparticles retained their transfection potency for more than 4 weeks (Mao *et al.*, 1997). Chitosan was also reported to have an immune stimulating activity such as increasing the accumulation and activation of macrophages and polymorphonuclear cells, promoting resistance to infections by cytokines, and enhancing CTL response (Seferian and Martinez, 2000; Kumar *et al.*, 2002). Studies also demonstrate that chitosan could not only slow down mucociliary clearance of encapsulated drugs but could also transiently increase paracellular absorption so as to improve immune stimulation (El-Shafy *et al.*, 2000). As both mucosal absorption enhancer and immune stimulator, chitosan represents a good mucosal delivering vehicle for DNA or protein vaccines. Nanoparticles made from high-molecular weight chitosan and pCMVArah2 plasmid DNA encoding for a major peanut allergen elicited secretory IgA and serum IgG2a production in mice after oral administration (Roy *et al.*, 1999). Challenge tests showed less severe and delayed anaphylactic responses in sensitized mice immunized with chitosan/pCMVArah2, than in the control group (non-treated animals) or in mice treated with chitosan /DNA lacking the Arah2 gene. Also, it has been reported that intragastric priming with GRA 1 protein vaccine loaded chitosan nanoparticles and boosting with GRA1pDNA vaccine resulted in high anti-GRA 1 antibodies, as characterized by a mixed IgG2a/IgG1 ratio (Bivas-Benita *et al.*, 2003).

Chitosan and positively charged chitosan complexes can bind strongly to negatively charged sites in the nasal cavity such as the sialic acid residues in the mucin. Furthermore, chitosan has in Caco-2 cell studies been shown to transiently open the tight junctions between cells thereby allowing the increased membrane transport of large molecular weight drugs (Illum, 1998) and to a lesser degree particulate systems coated with chitosan (Brooking *et al.*, 2001). Hence, although the exact mechanism of action of nasally administered plasmid DNA–chitosan vaccine system is not known, it was proposed that the small plasmid DNA–chitosan nanoparticles to some extent are able to pass the membrane and reach the underlying lymphoid tissue where they are transfected within antigen presenting cells. There might also be a possibility that the nanoparticles are taken up by the M-cell like cells in the NALT and presented to the underlying lymphoid tissue (Illum *et al.*, 2001).

Intranasal administration with chitosan-DNA complex resulted in transgenic DNA expression in mouse nasopharynx to induce Coxsackievirus B3 (CVB3) specific immune responses. Mice immunized with chitosan–DNA (pcDNA3-VP1) encoding VP1, major structural protein of CVB3, produced much higher levels of serum IgG and mucosal secretory IgA compared to mice treated with pcDNA3-VP1 or pcDNA3. Increased virus specific cytotoxic activity of spleen cells derived from chitosan–DNA vaccinated mice was also determined. Chitosan-pcDNA3-VP1 intranasal immunization resulted in 42.9% protection of mice against lethal CVB3 challenge and a significant reduction of viral load after acute CVB3 infection. Meanwhile no myonecrosis or infiltrating immune cells indicating ongoing myocarditis was detected in hearts of surviving mice treated with chitosan-DNA (Xu *et al.*, 2004). Recently, our group also developed chitosan nanoparticles loaded with plasmid DNA encoding surface protein of Hepatitis B virus and evaluated for their ability to induce humoral mucosal and cellular immune responses following intranasal immunization (Khatri *et al.*, 2008b). Results of the study showed that chitosan–DNA intranasal immunization induced both systemic and mucosal humoral immune responses (Figure 7).

Further, significant levels of both IL-2 and IFN-γ (Figure 8) were measured in mice immunized with pDNA administered IM and with pDNA loaded chitosan nanoparticles administered IN as compared to those recorded for control and alum adsorbed HBsAg. Both Th1 dependent cytokines and their high levels are evidenced for the strong cell-mediated immune response elicited by pDNA loaded chitosan formulations administered intranasally.

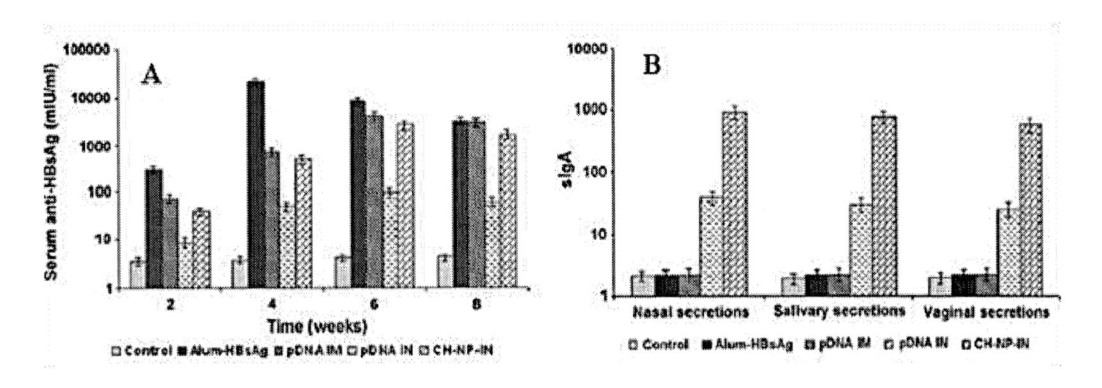


Figure 7. Serum anti-HBSAg profile (A) and secretory IgA levels in nasal, salivary and vaginal secretions (B), of mice immunized with various formulations. From Khatri *et al.*, 2008b.

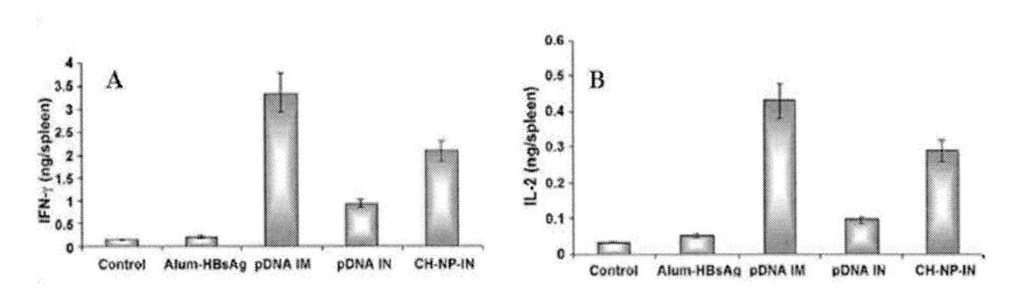


Figure 8. IFN-γ (C) and IL-2 (D) levels in spleen, of mice immunized with various formulations. From Khatri et al., 2008b.

Pulmonary administration of the pDNA encoding eight HLA-A*0201-restricted T-cell epitopes from *Mycobacterium tuberculosis* incorporated in chitosan nanoparticles was shown to induce maturation of dendritic cells (DCs), increased levels of IFN-γ secretion compared to pulmonary delivery of plasmid in solution or the more frequently used IM immunization route (Bivas-Benita *et al.*, 2004). Chitosan-based nanoparticles containing pDNA appears to be a promising approach for topical genetic immunization. Several different chitosan-based nanoparticles i.e., pDNA-condensed chitosan nanoparticles, and surface modified i.e., pDNA-coated on preformed cationic chitosan/carboxymethylcellulose (CMC) nanoparticles containing pDNA resulted in both detectable and quantifiable levels of luciferase expression in mouse skin 24 h after topical application, and significant antigen-specific IgG titer to expressed β-galactosidase at 28 days (Cui and Mumper, 2001). Except for the chitosan oligomer/CMC (300:100 w/w) nanoparticles, all other chitosan-based formulations resulted in IgG titers that were not statistically different than naked pDNA. The IgG titer was up to 32-fold greater when mice were immunized with pDNA coated on chitosan oligomer/CMC nanoparticles as compared to those mice immunized with naked pDNA alone.

#### *4.2.4. Dendrimers*

Another promising approach for polymeric gene delivery is the use of dendrimers. Dendrimers are repeatedly branched cascade polymers that are synthesized from a central core and branch outward. With each increasing synthesis generation, dendrimers become more branched, larger in size, and have a multiplicative increase in the number of end groups (Lee *et al.*, 2005). Like PEI, several dendrimers including poly(amidoamine) (PAMAM), make effective gene delivery polymers due to their large number of secondary and tertiary amines that can buffer the endosomes (Wood *et al.*, 2005). The complexation process between dendrimers and nucleic acids does not seem to differ fundamentally from other cationic polymers with high charge density: dendrimers interact with various forms of nucleic acids, such as plasmid DNA or antisense oligonucleotides, to form complexes which protect the nucleic acid from degradation. The interaction between dendrimer and nucleic acids is based on electrostatic interactions and lacks any sequence specificity (Figure 9). During the complexation the extended configuration of plasmid DNA is changed and a more compact configuration achieved, with the cationic dendrimer amines and the anionic phosphate group of nucleic acid reaching local charge neutralisation and the formation of DNA-dendrimer complexes (dendriplexes) (Bielinska *et al.*, 1997; Tang and Szoka, 1997; Chen *et al.*, 2000).

Particular advantages of dendrimers include a more monodisperse population than other polymers, which are typically polydisperse, and fine control over dendrimer generation number that can be tuned for a desired application thereby controlling dendrimer size, number of functional groups, etc. A disadvantage of dendrimers is that they require a multi-step synthesis that can be more expensive and time-consuming than the synthesis of linear polymers. Dendrimers do not need to be in their perfect spherical shape to function. In fact, efficacy can be improved by effectively pruning the branches of the dendrimer or by using the branches themselves. Partially degraded or ''fractured'' PAMAM dendrimers have been shown to have dramatic >50-fold enhancement for gene delivery as compared to complete PAMAM (Tang *et al.*, 1996). An interesting recent approach to build on dendrimers technology is the development of hybrid linear dendrons for targeted gene delivery. Wood *et al.*, (2005) have developed a self-assembling PAMAM dendrimer PEG linear block copolymer for targeting APCs.

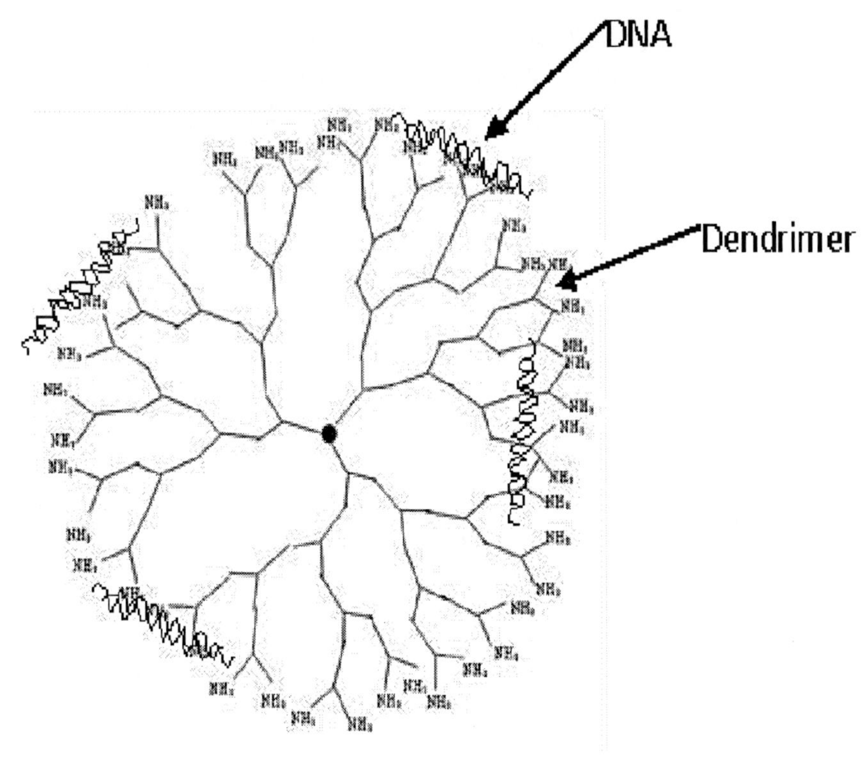


Figure 9. Dendrimer-DNA complex.

The linear-dendrimer block copolymer can be functionalized for APC-specific targeting by conjugation of a mannose receptor ligand at the end of the PEG chain. This multi-functional platform system was rationally designed to efficiently deliver pDNA for DNA vaccination and exhibits specific structure-based functionalities for bypassing many of the barriers of intracellular and extracellular nucleic acid delivery. The PAMAM block of the copolymer enhances transfection while the PEG linker prevents non-specific adherence to cell membranes (Kircheis *et al.*, 2001). Mannose targeting increases APC transfection, and engagement of these receptors may also stimulate APC activation (Gordon, 2002). The PAMAM–PEG–mannose block copolymer demonstrated effective condensation of DNA into organized, uniform nanoparticles approximately 150 nm in size. *In vitro* these NPs exhibited low cytotoxicity and high ligand-specific transfection efficiencies when compared to commercially available PEI, particularly the fifth and sixth generation dendrimers.

In a recent study, dendriplexes, complexes of dendrons and condensed plasmids containing the gene for protective antigen (PA) of Bacillus anthracis, were encapsulated in poly-lactide-co-glycolide (PLGA) particles using the double emulsion method. The two dendrons employed were with three C(18) chains (C(18) dendron) and one with no attached hydrocarbon chains (the C(0) dendron). Three types of particles were examined, namely PLGA-C(18) dendriplexes, PLGA-C(0) dendriplexes and the control PLGA-naked DNA system. Antibody titres were measured following three IM immunizations using 14 μg of DNA per dose at weekly intervals in BALB/c mice and results indicate that the PLGA-C(18) dendriplex particles produced superior levels of anti-PA IgG antibodies in comparison to animals immunized with the PLGA-C(0) dendriplex particles (Ribeiro *et al.*, 2007).

#### *4.2.5. Poly (Lactide-Co-Glycolide) (PLG)*

The encapsulation and release of protein antigen or plasmid DNA from microparticles has proven to be a very effective strategy for passively targeting vaccines to pro-APCs by size exclusion. Microparticles in the range of 1–10mm diameter are too large for endocytosis and therefore avoid general cellular uptake, but are small enough to be phagocytosed by MPs and DCs. Encapsulation of plasmid DNA also protects from nuclease degradation. Because of the bulk size of microparticles, a depot is formed at the injection site allowing for sustained exposure to DNA over time as microparticles are slowly cleared by phagocytosing cells (Nguyen *et al.*, 2008). Randolph *et al.*, (1999) showed that monocytes are recruited into the injection site within hours, where they phagocytose microparticles and differentiate into DCs by the time they have migrated to draining lymph nodes. Microencapsulation of DNA also provides a method of controlling release rates of DNA, which may be important for timing immune responses by coordinating DC migration to lymph nodes, maturation, and presentation of costimulatory molecules, peak gene expression, and antigen presentation.

The first attempt to deliver plasmid as a vaccine using PLGA microencapsulation was made by Hedley *et al.*, (1998) who showed that stronger CTL responses were elicited using microencapsulated plasmid delivery SC and IP compared to naked plasmid injections using a VSV antigen system. PLGA has been approved by FDA for several therapeutic applications and exhibits biodegradability, biocompatibility and safety in humans. Zycos has a therapeutic human papillomavirus vaccine (ZYC101a) in Phase 1 clinical trials (Klencke *et al.*, 2002), comprising plasmid encapsulated in PLGA microparticles. The plasmid encodes multiple HLA-restricted epitopes derived from human papilloma virus (HPV-16 and 18) E6 and E7 proteins, which are HPV oncoproteins consistently expressed in neoplastic cells (Garcia *et al.*, 2004). In preclinical mouse studies, formulation of PLGA with taurocholic acid (TA) or monomethoxyl polyethyleneglycol-distearoylphosphatidylethanolamine (PEG-DSPE) has been described. The exact mechanism of action of this formulation is unknown, but particulates and their components may be proinflammatory and/or may target the DNA to APCs (Mckeever *et al.*, 2002). Clinical trials using the PLGA microparticle delivery system showed that 83% of patients demonstrated an immune response that persisted for six months using a plasmid encoding (HPV-16 E7) (Klencke *et al.*, 2002). It was also demonstrated that oral immunization with plasmid DNA led to protective immunity against rotavirus challenge when PLGA microparticles were used for delivery (Chen *et al.*, 1998). Also, plasmid DNA vaccines developed by Chiron Corporation encoding human immunodeficiency virus (HIV) Gag and Env adsorbed onto the surface of cationic PLGA microparticles were shown to be substantially more potent than corresponding naked DNA vaccines. The magnitude of anti-Env antibody responses induced by PLGA/DNA particles was equivalent to that induced by recombinant gp120 protein formulated with a strong adjuvant, MF-59 (O'Hagan *et al.*, 2001).

The attractiveness of PLGA has generated a great deal of interest into the advancement of this technology to the clinic. Although studies demonstrate the ability of PLGA to function as a genetic vaccine delivery vehicle, this polymer was never designed for this particular application and has several limiting disadvantages. Acidic degradation products that build up in the microparticle interior can severely stunt or permanently damage the activity of plasmid DNA. This can be attributed to PLGA ester bond degradation leading to acids that cannot easily diffuse out and away from the particle interior. It has been demonstrated using pH sensitive fluorescent probes and microscopy that the pH can drop to as low as 2 after three days of incubation (Fu *et al.*, 2000). Although this internal pH microclimate can stabilize some drugs (Shenderova *et al.*, 1999), low pH has been shown to completely abolish plasmid transfection activity below a pH of 4. In addition, the amount of time needed for quantitative release of plasmid DNA from even low molecular weight PLGA microparticles is on the order of two weeks, while the lifespan of the majority of DCs after activation is approximately ten days.

Encapsulation-based approaches are often associated with instability and degradation of the entrapped molecules occuring both during the encapsulation and release processes. To overcome these problems, physical adsorption on the surface of nano-and micro-particles has very recently been proposed and shown to be a simpler, more efficient and less damaging procedure for the antigen. The reversible adsorption of the antigen to the surface of the nano- and micro-particles has been achieved with the addition of surfactants and/or detergents during particle preparation. Surface adsorbed DNA based particulate carriers not only increase the loading efficiency but also reduce the degradation of DNA during formulation steps. Additionally, it also avoids the difficulties associated with slower erosion-dependent release kinetics, and may increase transfection efficiency by making DNA immediately available at the surface. Materials that have been used for facilitating surface loading of DNA were cationic surfactant cetyltrimethylammonium bromide (CTAB), PEI, DOTAP etc. These materials can be conjugated at the surface of preformed particles or can be mixed during preparation of particles to provide cationic charged to the polymeric particles.

Adsorbing plasmid DNA onto PLGA/CTAB microspheres has been researched by the Chiron Corporation (now Novartis Vaccines) for IM vaccination. Surface adsorption of plasmid DNA results in high DNA loading (effectively 100% adsorption of DNA). Further, it also increases transfection of DCs and antigen presentation *in vitro* (Denis-Mize *et al.*, 2000). *In vivo* the PLGA/CTAB system resulted in increased transfection of muscle tissue following IM injection, increased production of antibody titers, and greater activation of CTL responses to plasmid-encoded HIV antigens when compared to PLGA microparticles or naked DNA (IM) alone (Singh *et al.*, 2000).

Incorporating PEI into PLGA microspheres has also been developed as a method for avoiding the problems associated with internal encapsulation of plasmid DNA (Oster *et al.*, 2005). PEI imparts a positive charge to the PLGA microsphere. Unlike CTAB, which by itself is not a transfection agent, PEI has intrinsic ability to form nanoparticles with DNA and increase transfection efficiency. Kasturi *et al.* (2005) have reported a method for covalently attaching PEI to the PLGA microparticle surface that minimizes the amount of labile PEI, thereby incorporating the endosomal buffering capacity of PEI without high toxicity. PLGA–PEI microspheres prepared by this method improve *in vitro* transfection and cause upregulation of costimulatory signals on APCs. Branched polyethyleneimine (b-PEI) was covalently conjugated to the surface of PLGA microparticles using carbodiimide chemistry to create cationic microparticles capable of simultaneously delivering both DNA vaccines as well as other immunomodulatory agents (cytokines or nucleic acids) within a single injectable delivery vehicle. Covalent conjugation of b-PEI allows efficient surface loading of nucleic acids, introduces intrinsic buffering properties to PLGA particles and enhances transfection of phagocytic cells without affecting the cytocompatibility of PLGA carriers. Furthermore, a pH-sensitive, degradable poly-β-aminoester (PBAE) in tandem with low molecular weight PLGA has been studied as a DNA vaccine delivery system (Little *et al.*, 2004). This type of particles can quantitatively release encapsulated material in the range of endosomal pH due to a unique balance of hydrophobicity and charge-inducible tertiary amines. These tertiary amines in the PBAE backbone may also act as a weak base (or proton sponge), which may mediate phagosomal/lysosomal release. These amine groups also act to buffer the internal aqueous environment due to ester bond degradation, and plasmid extracted from these particles after incubation in aqueous media has been shown to have higher integrity than pure PLGA microparticles. These hybrid PBAE/PLGA microparticles have exhibited enhanced delivery of plasmid DNA to APCs when compared with PLGA alone and have strong adjuvant effects on DCs *in vitro*.

## 5. Concluding Remark and Future Prospects

There is a great potential for the use of nanoparticulate systems in DNA vaccine development; it remains to be seen, however, if that potential is realised. DNA vaccines can circumvent many of the problems associated with recombinant protein-based vaccines, such as high costs of production, difficulties in purification, incorrect folding of antigen and poor induction of CD8+ T cells. DNA also has clear advantages over recombinant viruses, which are plagued with the problems of pre-existing immunity, risk of insertion mutagenesis, loss of attenuation or spread of inadvertent infection. Perhaps, the primary goal of genetic vaccines should not be to replace well established conventional vaccines with a good track record, instead the focus should be on diseases for which conventional vaccine approaches are ineffective. Furthermore, the possibility of inducing immunotolerance or autoimmune diseases also needs to be investigated more thoroughly, in order to arrive at a well-founded consensus, which justifies the widespread application of DNA vaccines in a healthy population.

## References

Adamina M, Guller U, Bracci L, Heberer M, Spagnoli GC, Schumacher R. Clinical applications of virosomes in cancer immunotherapy. *Expert Opin. Biol. Ther.* 2006; 6: 1113-1121.

Ahmed OAA, Adjimatera N, Pourzand C, Blagbrought IS. $N^4$, $N^9$-Dioleolyl spermine is a novel nonviral lipopolyamine vector for plasmid DNA formulation. *Pharm. Res.,* 2005; 22: 972-980.

Azzam T, Eliyahu H, Makovitzki A, Linial M, Domb AJ. Hydrophobized dextran-spermine conjugates as potential vector for *in vitro* gene transfection. *J. Control. Rel.,* 2004; 96: 309-323.

Benns JM, Choi JS, Mahato RI, Park JS, Kim SW. pH-sensitive cationic polymer gene delivery vehicle: N-Ac-poly(L-histidine)-graft-poly(L-lysine) comb shaped polymer. *Bioconjug. Chem.* 2000; 11: 637-645.

Bielinska AU, Kukowska-Latallo JF, Baker Jr. JR. The interaction of plasmid DNA with polyamidoamine dendrimers: mechanism of complex formation and analysis of alterations induced in nuclease sensitivity and transcriptional activity of the complexed DNA. *Biochim. Biophys. Acta* 1997; 1353: 180-190.

Bivas-Benita M, Laloup M, Versteyhe S *et al.* Generation of *Toxoplasma gondii* GRA1 protein and DNA vaccine loaded chitosan particles: preparation, characterization, and preliminary *in vivo* studies. *Int. J. Pharm.*, 2003; 266: 17-27.

Bivas-Benita M, van Meijgaarden KE, Franken KL, *et al.* Pulmonary delivery of chitosan-DNA nanoparticles enhances the immunogenicity of a DNA vaccine encoding HLA-A*0201-restricted T-cell epitopes of Mycobacterium tuberculosis. *Vaccine*, 2004; 22: 1609-1615.

Bivas-Benita M, Bar L, Gillard GO, *et al.* Efficient Generation of Mucosal and Systemic Antigen-Specific CD8+ T-Cell Responses following Pulmonary DNA Immunization. *J. Virol.*, 2010; 84: 5764-5774.

Bivas-Benita M. van Meijgaarde KE, Franken KLMC, *et al.* Pulmonary delivery of chitosan-DNA nanoparticles enhances the immunogenicity of a DNA vaccine encoding HLA-A*0201-restricted T-cell epitopes of *Mycobacterium tuberculosis. Vaccine* 2004; 22: 1609-1615.

Boussif O. Lezoualch, F, Zanta MA, *et al.* A versatile vector for gene and oligonucleotide transfer into cells in culture and *in vivo*: polyethylenimine. *Proc. Natl. Acad. Sci. USA* 1995; 92: 7297-7301.

Bramwell VW. Eyles JE, Somavarapu S, Alpar HO. Liposome/DNA complexes coated with biodegradable PLA improve immune responses to plasmid encoding hepatitis B surface antigen. Immunology 2002; 106: 412-418.

Brooking J, Davis SS, Illum L. Transport of nanoparticles across the rat nasal mucosa. *J. Drug. Target*, 2001; 9: 267-279.

Brown MD, Schatzlein AG, Uchegbu IF. Gene delivery with synthetic (nonviral) carriers. *Int. J. Pharm.* 2001; 229: 1-21.

Chen SC, Jones DH, Fynan EF, *et al.* Protective immunity induced by oral immunization with a rotavirus DNA vaccine encapsulated in microparticles. *J. Virol.* 1998; 72: 5757-5761.

Chen W, Turro NJ, Tomalia DA. Using ethidium bromide to probe the interactions between DNA and dendrimers. *Langmuir* 2000; 16: 15-19.

Chiu SJ, Marcucci G, Lee RJ. Efficient delivery of an antisense oligodeoxyribonucleotide formulated in folate receptor-targeted liposomes. *Anticancer. Res.,* 2006; 26: 1049-1056.

Cho YW, Kim JD, Park KJ. Polycation gene delivery systems: escape from endosomes to cytosol. *J. Pharm. Pharmacol.,* 2003; 55: 721-734.

Correale P, Cusi MG, Sabatino M, *et al.* Tumour-associated antigen (TAA)-specific cytotoxic T cell (CTL) response *in vitro* and in a mouse model, induced by TAA-plasmids delivered by influenza virosomes. *Eur. J. Cancer* 2001; 37: 2097-2103.

Cui Z, Mumper RJ. Chitosan-based nanoparticles for topical genetic immunization. *J. Control. Rel.* 2001; 75: 409-419.

Cusi MG, Del Vecchio MT, Terrosi C *et al.*: Immune-reconstituted influenza virosome containing CD40L gene enhances the immunological and protective activity of a carcinoembryonic antigen anticancer vaccine. *J. Immunol.,* 2005; 174:7210-7216.

Cusi MG, Terrosi C, Savellini GG, Genova GD, Zurbriggen R, Correale P. Efficient delivery of DNA to dendritic cells mediated by influenza virosomes. *Vaccine,*2004; 22:736-740.

Cusi MG. Zurbriggen R, Valassina M, *et al.* Intranasal immunization with mumps virus DNA vaccine delivered by influenza virosomes elicits mucosal and systemic immunity. *Virology* 2000; 277: 111-118.

Denis-Mize KS, Dupuis M, MacKichan ML, Singh M, Doe B, O'Hagan D, Ulmer JB, Donnelly JJ, McDonald DM, Ott G. Plasmid DNA adsorbed onto cationic microparticles mediates target gene expression and antigen presentation by dendritic cells. *Gene Ther.,* 2000; 7: 2105-2112.

Densmore CL, Orson FM, Xu B, *et al.* Aerosol delivery of robust polyethyleneimine-DNA complexes for gene therapy and genetic immunization. *Mol. Ther.* 2000; 1: 180-188.

Dzau VJ, Mann MJ, Morishita R, Kaneda Y. Fusigenic viral liposome for gene therapy in cardiovascular diseases. *Proc. Natl. Acad. Sci. USA* 1996; 93: 11421-11425.

El-Shafy MA, Kellaway IW, Taylor G, Dickinson PA. Improved nasal bioavailability of FITC-dextran (Mw4300) from mucoadhesive microspheres in rabbits. *J. Drug Target,* 2000; 7: 355-361.

Fajac I, Allo JC, Souil E, *et al.* Histidylated polylysine as a synthetic vector for gene transfer into immortalized cystic fibrosis airway surface and airway gland serous cells. *J. Gene Med.* 2000; 2: 368-378.

Felgner PL, Gadek TR, Holm M, Roman R, Chan HW, Wenz M, Northrop JP. Lipofection: a highly efficient, lipid-mediated DNA-transfection procedure. *Proc. Natl. Acad. Sci. USA,* 1987; 84: 7413-7417.

Felnerova D, Viret JF, Gluck R, Moser C. Liposomes and virosomes as delivery systems for antigens, nucleic acids and drugs. *Current Opin. in Biotechnology,* 2004; 15: 518–529.

Fischer D, Bieber T, Li Y, Elsässer H, Kissel T. A novel non–viral vector for DNA delivery based on low molecular weight, branched polyethylenimine: Effect of molecular weight on transfection efficiency and cytotoxicity. *Pharm. Res.,* 1999; 16: 1273-1279.

Forrest ML, Meister GE, Koerber JT, Pack DW. Partial acetylation of polyethylenimine enhances *in vitro* gene delivery. *Pharm. Res.* 2004; 21: 365-371.

Fu K, Pack DW, Klibanov AM, Langer R. Visual evidence of acidic environment within degrading poly(lactic-co-glycolic acid) (PLGA) microspheres. *Pharm. Res.* 2000; 17: 100-106.

Garcia F, Petry KU, Muderspach L, *et al.* ZYC101a for treatment of highgrade cervical intraepithelial neoplasia: a randomized controlled trial. *Obstet. Gynecol.* 2004; 103: 317-326.

Garzon MR, Berraondo P, Crettaz J, *et al.* Induction of gp120-specific protective immune responses by genetic vaccination with linear polyethylenimine-plasmid complex. *Vaccine* 2005; 23: 1384-1392.

Godbey WT, Wu KK, Mikos AG. Poly(ethylenimine) and its role in gene delivery. *J. Control Rel.,* 1999; 60: 149-160.

Gordon S. Pattern recognition receptors: doubling up for the innate immune response. *Cell,* 2002; 111: 927-930.

Gregoriadis G, Mccormack B, Obrenovic M, Saffie R, Zadi B, Perrie Y. Vaccine entrapment in liposomes. *Methods,* 1999; 19:156-162.

Gregoriadis G. Bacon A, Caparros-Wanderley W, McCormack B. A role for liposomes in genetic vaccination. *Vaccine* 2002; 20 Suppl 5: B1-9.

Gregoriadis G. Liposomes in the therapy of lysosomal storage diseases. *Nature,* 1978; 275: 695–696.

Gregoriadis G. Saffie R, de Souza JB. Liposome-mediated DNA vaccination. *FEBS Lett.* 1997; 402: 107-110.

Gustafsson J, Arvidson G, Karlsson G, Almgren M. Complexes between cationic liposomes and DNA visualized by cryo-TEM. *Biochim. Biophys. Acta*, 1995; 1235: 305–312.

Haensler J, Szoka FC Jr. Polyamidoamine cascade polymers mediate efficient transfection of cells in culture. *Bioconjugate Chem.,* 1993; 4: 372-379.

Hattori Y. Kawakami S, Suzuki S, Yamashita F, Hashida M. Enhancement of immune responses by DNA vaccination through targeted gene delivery using mannosylated cationic liposome formulations following intravenous administration in mice. *Biochem. Biophys. Res. Commun.* 2004; 317: 992-999.

Hedley ML, Curley J, Urban R. Microspheres containing plasmid-encoded antigens elicit cytotoxic T-cell responses. *Nat. Med.* 1998; 4: 365-368.

Hirano T, Fujimoto HJ, Ueki T, *et al.* Persistent gene expression in rat liver *in vivo* by repetitive transfection using HVJ-liposome. *Gene Ther.*, 1998; 5: 459-464.

Huang M., Fong C-W, Khor E, Lim L-Y. Transfection efficiency of chitosan vectors: Effect of polymer molecular weight and degree of deacetylation. *J. Control. Rel.* 2005; 106: 391-406.

Illum L, Jabbal-Gill I, Hinchcliffe M, Fisher AN, Davis SS. Chitosan as a novel nasal delivery system for vaccines. *Adv. Drug. Deliv. Rev.,* 2001; 51: 81-96.

Illum L. Chitosan and its use as a pharmaceutical excipient. *Pharm. Res.,* 1998; 15: 1326-1331.

Iqbal M, Lin W, Jabbal-Gill I, Davis SS, Steward MW, Illum L. Nasal delivery of chitosan-DNA plasmid expressing epitopes of respiratory syncytial virus (RSV) induces protective CTL responses in BALB/c mice. *Vaccine,* 2003; 21: 1478-1485.

Jiao X. Wang RYH, Feng Z, Alter HJ, Shih JWK. Modulation of Cellular Immune Response Against Hepatitis C Virus Nonstructural Protein 3 by Cationic Liposome Encapsulated DNA Immunization. *Hepatology* 2003; 37: 452-460.

Kamiya H, Dugue L, Yakushiji H, Pochet S, Nakabeppu Y, Harashima H. Substrate recognition by the human MTH1 protein. Nucleic Acids Res Suppl, 2002; 2: 85-86.

Kaneda Y, Saeki R, Marishita R. Gene therapy using HVJ-liposomes: the best of both worlds. *Mol. Med. Today*, 1999; 5: 298-303.

Kasturi SP, Sachaphibulkij K, Roy K. Covalent conjugation of polyethyleneimine on biodegradable microparticles for delivery of plasmid DNA vaccines. *Biomaterials* 2005; 26: 6375-6385.

Khatri K, Goyal AK, Gupta PN, Mishra N, Mehta A, Vyas SP. Surface modified liposomes for nasal delivery of DNA vaccine. *Vaccine,* 2008a; 26: 2225-2233.

Khatri K, Goyal AK, Gupta PN, Mishra N, Vyas SP. Plasmid DNA loaded chitosan nanoparticles for nasal mucosal immunization against hepatitis B. *Int. J. Pharm.,* 2008; 354: 235-241.

Khatri K, Goyal AK, Vyas SP. Potential of nanocarriers in genetic immunization. *Recent Pat. Drug Deliv. Formul.,* 2008c; 2: 68-82.

Kircheis R, Wightman L, Wagner E. Design and gene delivery activity of modified polyethylenimines. *Adv. Drug Deliv. Rev.,* 2001; 53: 341-358.

Klemm AR, Young D, Lloyd JB. Effects of poly-ethylenimine on endocytosis and lysosome stability. *Biochem. Pharmacol.* 1998; 56: 41-46.

Klencke B, Matijevic M, Urban RG, *et al.* Encapsulated plasmid DNA treatment for human papillomavirus 16-associated anal dysplasia: a Phase I study of ZYC101. *Clin. Cancer Res.* 2002; 8: 1028-1037.

Konstan MW, Davis PB, Wagener JS, *et al.* Compacted DNA nanoparticles administered to the nasal mucosa of cystic fibrosis subjects are safe and demonstrate partial to complete cystic fibrosis transmembrane regulator reconstitution. *Hum. Gene Ther.* 2004; 15: 1255-1269.

Kumar M, Behera AK, Lockey RF, Zhang J, Bhullar G, De La Cruz CP, Chen LC, Leong KW, Huang SK, Mohapatra SS. Intranasal gene transfer by chitosan DNA nanospheres protects Balb/c mice against acute respiratory syncytial virus infection. *Hum. Gene Ther.,* 2002; 13: 1415-1425.

Lambert RC, Maulet Y, Dupont J-L *et al.* Polyethylenimine-mediated DNA trans fection of peripheral and central neurons in primary culture: 21 probing Ca channel structure and function with antisense oligonucleotides. *Mol. Cell Neurosci.,* 1996; 7: 239-246.

Lechardeur D, Verkman AS, Lukaes GL. Intracellular routing of plasmid DNA during non-viral gene transfer. *Adv. Drug. Del. Rev*., 2005; 57: 755-767.

Lee CC, MacKay JΛ, Frechet JM, Szoka FC. Designing dendrimers for biological applications. *Nat. Biotechnol.,* 2005; 23: 1517-26.

Li XW, Lee DKL, Chan ASC, Alpar HO. Sustained expression in mammalian cells with DNA complexed with chitosan nanoparticles. *Biochim. Biophys. Acta* 2003; 1630: 7-18.

Little SR, Lynn DM, Ge Q, *et al.* Poly-beta amino ester-containing microparticles enhance the activity of nonviral genetic vaccines. *Proc. Natl. Acad. Sci. USA* 2004; 101: 9534-9539.

Liu G, Li D, Pasumarthy MK, *et al.* Nanoparticles of compacted DNA transfect postmitotic cells. *J. Biol. Chem.* 2003; 278: 32578-32586.

Lu Y, Kawakami S, Yamashita F, Hashida M. Development of an antigenpresenting cell-targeted DNA vaccine against melanoma by mannosylated liposomes. *Biomaterials.*2007; 28: 3255-3262.

Mahor S. Rawat A, Dubey PK, *et al.* Cationic transfersomes based topical genetic vaccine against hepatitis B. *Int. J. Pharm.* 2007; 340: 13-19.

Mao HQ, Troung-Le VL, August JT, Leong KW. DNA-chitosan nanospheres: derivatization and storage stability. *Proc. Intl. Symp. Control Rel. Bioact. Mater.* 1997; 24: 671-672.

Marten A, Ziske C, Schottker B, *et al.* Transfection of dendritic cells (DCs) with the CIITA gene: increase in immunostimulatory activity of DCs. *Cancer Gene Ther.* 2001; 8: 211-219.

McKeever U, Barman S, Hao T, *et al.* Protective immune responses elicited in mice by immunization with formulations of poly(lactide-co-glycolide) microparticles. *Vaccine* 2002; 20: 1524-1531.

McKenzie DL, Smiley E, Kwok KY, Rice KG. Low molecular weight disulfide cross–linking peptides as nonviral gene delivery carriers. *Bioconjug. Chem.,* 2000; 11: 901-909.

Midoux P, Kichler A, Boutin V, Maurizot JC, Monsigny M. Membrane permeabilization and efficient gene transfer by a peptide containing several histidines. *Bioconjugate Chem.,* 1998; 9: 260-267.

Midoux P, Monsigny M. Efficient gene transfer by histidylated polylysine/pDNA complexes. *Bioconjug. Chem.* 1999; 10: 406-411.

Minigo G, Scholzen A, Tang CK, Hanley JC, Kalkanidis M, Pietersz GA, Apostolopoulos V, Plebanski M. Poly-l-lysine-coated nanoparticles: A potent delivery system to enhance DNA vaccine efficacy. *Vaccine,* 2007; 25: 1316-1327.

Miyata K. Fukushima S, Nishiyama N, Yamasaki Y, Kataoka K. PEG-based block catiomers possessing DNA anchoring and endosomal escaping functions to form polyplex micelles with improved stability and high transfection efficacy. *J. Control Rel.* 2007; 122: 252-260.

Nakanishi M, Noguchi A. Confocal and probe microscopy to study gene transfection mediated by cationic liposomes with a cationic cholesterol derivative. *Adv. Drug Del.* Rev, 2001; 52: 197-207.

O'Hagan D, Singh M, Ugozzoli M, *et al.* Induction of potent immune responses by cationic microparticles with adsorbed human immunodeficiency virus DNA vaccines. *J. Virol.* 2001; 75: 9037-9043.

Okada M, Kita Y, Nakajima T, *et al.* Evaluation of a novel vaccine (HVJliposome/HSP65 DNA+ IL-12 DNA) against tuberculosis using the cynomolgus monkey model of TB. *Vaccine,* 2007; 25: 2990-2993.

Okamoto T, Kaneda Y, Yuzuki D, Huang SK, Chi DD, Hoon DS. Induction of antibody response to human tumor antigens by gene therapy using a fusigenic viral liposome vaccine. *Gene Ther.,* 1997; 4: 969-976.

Oku N, Yamaguchi N, Yamaguchi N, Shibamoto S, Ito F, Nango M. The fusogenic effect of synthetic polycations on negatively charged lipid bilayers. *J. Biochem.* 1986; 100: 935-944.

Orson FM, Kinsey BM, Hua PJ, Bhogal BS, Densmore CL, Barry MA. Genetic immunization with lung-targeting macroaggregated polyethyleneimine-albumin conjugates elicits combined systemic and mucosal immune responses. *J. Immunol.* 2000; 164: 6313-6321.

Oster CG, Kim N, Grode L, Barbu-Tudoran L, Schaper AK, Kaufmann SHE, Kissel T. Cationic microparticles consisting of poly(lactide-co-glycolide) and polyethylenimine as carriers systems for parental DNA vaccination. *J. Control Rel.,* 2005; 104: 359-377.

Patil SD, Rhodes DG, Burgess DJ. DNA-based therapeutics and DNA delivery systems: A comprehensive review. *The AAPS J.,* 2005; 7: E61-E77.

Perrie Y, Gregoriadis G. Liposome-entrapped plasmid DNA: characterisation studies. *Biochim. Biophys. Acta* (BBA) – General Subjects, 2000; 1475:125-132.

Perrie Y. Barralet JE, McNeil S, Vangala A. Surfactant vesicle-mediated delivery of DNA vaccines via the subcutaneous route. *Int. J. Pharm.* 2004; 284: 31-41.

Perrie Y. Frederik PM, Gregoriadis G. Liposome-mediated DNA vaccination: the effect of vesicle composition. *Vaccine* 2001; 19: 3301-3310.

Perrie Y. Obrenovic M, McCarthy D, Gregoriadis G. Liposome (Lipodine)-mediated DNA vaccination by the oral route. *J. Liposome Res.,* 2002; 12: 185-197.

Pouton CW, Lucas P, Thomas BJ, Uduehi AN, Milroy DA, Moss SH. Polycation-DNA complexes for gene delivery: a comparison of the biopharmaceutical properties of cationic polypeptides and cationic lipids. *J. Control Rel.,* 1998; 53: 289-299.

Randolph GJ, Inaba K, Robbiani DF, Steinman RM, Muller WA. Differentiation of phagocytic monocytes into lymph node dendritic cells *in vivo*. *Immunity*, 1999; 11: 753-761.

Rawat A, Vaidya B, Khatri K, Goyal AK, Gupta PN, Mahor S, Paliwal R, Rai S, Vyas SP.Targeted intracellular delivery of therapeutics: An overview. *Die Pharmazie,* 2007; 62:643-658.

Reddy JA, Abburi C, Hofland H, Howard SJ, Vlahov I, Wils P, Leamon CP. Folate-targeted, cationic liposome-mediated gene transfer into disseminated peritoneal tumors. *Gene Ther.,* 2002; 9:1542-1550.

Reese TA, Liang HE, Tager AM, Luster AD, Van Rooijen N, Voehringer D, Locksley RM. Chitin induces accumulation in tissue of innate immune cells associated with allergy. *Nature*, 2007; 447: 92-96.

Ribeiro S, Rijpkema SG, Durrani Z, Florence AT. PLGA-dendron nanoparticles enhance immunogenicity but not lethal antibody production of a DNA vaccine against anthrax in mice. *Int. J. Pharm.* 2007; 331: 228-232.

Roy K, Mao HQ, Huang SK, Leong KW. Oral gene delivery with chitosan-DNA nanoparticles generates immunologic protection in a murine model of peanut allergy, *Nat. Med.*1999; 5: 387-391.

Scardino P, Correale H, Firat M, *et al. In vitro* study of the GC90/IRIV vaccine for immune response and autoimmunity into a novel humanised transgenic mouse. *Br. J. Cancer* 2003; 89: 199-205.

Seferian PG, Martinez ML. Immune stimulating activity of two new chitosan containing adjuvant formulations. *Vaccine,* 2000; 19: 661-668.

Shenderova A, Burke TG, Schwendeman SP. The acidic microclimate in poly(lactide-co-glycolide) microspheres stabilizes camptothecins. *Pharm. Res.* 1999; 16: 241-248.

Shimizu N, Chen J, Gamou S, Takayanagi A. Immunogene approach toward cancer therapy using erythrocyte growth factor receptor-mediated gene delivery. *Cancer Gene Ther.* 1996; 3: 113-120.

Singh M, Briones M, Ott G, O'Hagan D. Cationic microparticles: A potent delivery system for DNA vaccines. *Proc. Natl. Acad. Sci. USA,* 2000; 97: 811-816.

Somavarapu S. Bramwell VW, Alpar HO. Oral plasmid DNA delivery systems for genetic immunisation. *J. Drug Target* 2003; 11: 547-553.

Tang MX, Redemann CT, Szoka Jr FC. *In vitro* gene delivery by degraded polyamidoamine dendrimers. *Bioconjug. Chem.*, 1996; 7: 703-714.

Tang MX, Szoka FC. The influence of polymer structure on the interactions of cationic polymers with DNA and morphology of the resulting complexes. *Gene Ther.* 1997; 4: 823-832.

Thanou M, Florea BI, Langemeyer MW, Verhoef JC, Junginger HE. N-trimethylated chitosan chloride (TMC) improves the intestinal permeation of the peptide drug buserelin *in vitro* (Caco-2 cells) and *in vivo* (rats). *Pharm. Res.,* 2000; 17: 27-31.

Torrieri-Dramard L, Lambrecht B, Ferreira HL, Van den Berg T, Klatzmann D, Bellier B. Intranasal DNA Vaccination Induces Potent Mucosal and Systemic Immune Responses and Cross-protective Immunity Against Influenza Viruses. *Molecular Therapy*, 2010, | doi:10.1038/mt.2010.222.

Tyagi R, Sharma PK, Vyas SP, Mehta A. Various carrier system(s)-mediated genetic vaccination strategies against malaria. *Expert Rev. Vaccines,* 2008; 7: 499-520.

Van de Wetering P, Moret EE, Schuurmans-Nieuwenbroek NME, van Steenbergen MJ, Hennink WE. Structure-activity relationships of water-soluble cationic methacrylate/methacrylamide polymers for nonviral gene delivery. *Bioconjugate Chem.*, 1999; 10: 589-597.

Wagner E, Ogris M, Zauner W. Polylysine–based transfection systems utilizing receptor–mediated delivery. *Adv. Drug Deliv. Rev.,* 1998; 30: 97-113.

Wagner E, Plank C, Zatloukal K, Cotten M, Birnstiel ML. Influenza virus hemagglutinin HA-2 N-terminal fusogenic peptides augment gene transfer by transferrin-polylysine-DNA complexes: toward a synthetic virus-like genetransfer vehicle. *Proc. Natl. Acad. Sci. USA* 1992; 89: 7934-7938.

Wiethoff CM, Middaugh CR. Barriers to nonviral gene delivery. *J. Pharm. Sci.,* 2003; 92: 203-217.

Wolfert MA, Dash PR, Nazarova O, *et al.* Polyelectrolyte vectors for gene delivery: influence of cationic polymer on biophysical properties of complexes formed with DNA. *Bioconjug Chem.* 1999; 10: 993-1004.

Wood KC, Little SR, Langer R, Hammond PT. A family of hierarchically self-assembling linear-dendritic hybrid polymers for highly efficient targeted gene delivery. *Angew Chem. Int. Ed.,* 2005; 44: 6704.

Wrobel I, Collins D. Fusion of cationic liposomes with mammalian cells occurs after endocytosis. *Biochim. Biophys. Acta*, 1995; 1235: 296-304.

Xu W, Shen Y, Jiang Z, Wang Y, Chu Y, Xiong S. Intranasal delivery of chitosan-DNA vaccine generates mucosal SIgA and anti-CVB3 protection. *Vaccine* 2004; 22: 3603-3612.

Xu Y, Szoka FC. Mechanism of DNA release from cationic liposome/DNA complexes used in cell transfection. *Biochemistry*, 1996; 35: 5616-5623.

Yoshida S, Tanaka T, Kita Y, *et al.* DNA vaccine using hemagglutinating virus of Japan-liposome encapsulating combination encoding mycobacterial heat shock protein 65 and interleukin-12 confers protection against *Mycobacterium tuberculosis* by T cell activation. *Vaccine,* 2006; 24: 1191-1204.

Yoshikawa T. Imazu S, Gao JQ, *et al.* Augmentation of antigen-specific immune responses using DNA-fusogenic liposome vaccine. *Biochem. Biophy. Res. Commun.* 2004; 325: 500-505.

Ziady AG, Gedeon CR, Miller T, *et al.* Transfection of airway epithelium by stable PEGylated poly-L-lysine DNA nanoparticles *in vivo*. *Mol. Ther.,* 2003; 8: 936-947.

*Reviewed by*: Dr. K. C. Gupta
*Address*: Indian Institute of Toxicology Research
Post Box No. 80, Mahatma Gandhi Marg
Lucknow -226 001, India
Tel: +91-522-2621856, Fax : +91-522-2628227, 2611547
Email: kcgupta@iitr.res.in

In: DNA Vaccines: Types, Advantages and Limitations ISBN 978-1-61324-444-9
Editors: E. C. Donnelly and A. M. Dixon 

*Chapter II*

# Improving the Immunogenicity of DNA Vaccines: A Nano-Sized Task?

***Aya C. Taki,[*] Natalie Kikidopoulos[*], Fiona J. Baird and Peter M. Smooker[2]***
School of Applied Sciences, RMIT University, Bundoora 3083, Australia

## Abstract

A major biotechnological revolution of the last two decades has resulted in the novel application of DNA as a vaccine to prevent or treat disease. This has progressed from principle to product in only 20 years. However, the wider application of DNA vaccines is reliant on improving their immunogenicity, as while DNA vaccination can provide a broad immune response to an antigen, the magnitude is often less than optimal. This has led to many studies into improving the immunogenicity by using different delivery methods, targeting to specific cells in the host and by co-delivery with potent immune stimulators. Here we will review the different methodologies used to improve the immunogenicity of DNA vaccines, designed to maximise immunogenicity without changing the fundamental simplicity and usability of this type of vaccine. The most recent innovations, of delivery using nanoparticles, are also highlighted.

**Keywords:** immunogenicity; DNA vaccine; nanoparticle

## Introduction

DNA vaccination (also termed genetic immunization) is a technique that has over the past twenty years both raised hopes of frustrated vaccine researchers, and subsequently provided some disappointments. Although this is the nature of scientific discoveries, the hope (hype?)

---

* Equal contribution.

2 Corresponding author: peter.smooker@rmit.edu.au.

that surrounded the emerging technique in the early 1990's was almost unparalleled. The three key papers that ushered in the age of DNA vaccination started in 1990, where it was shown that expression of heterologous protein in myoctes after plasmid immunization continued for at least two months [1]. This was followed by the demonstration that the protein expression could induce immune responses [2], and the key paper that showed the generation of CTL's and heterologous protection against influenza in a mouse model [3]. The latter finding really set the field in motion, as the generation of CTL's is essential for vaccination against many pathogens (and indeed, cancer).

The induction of CTL's is one of the major advantages of DNA vaccines. The other is more practical-they are very simple to produce. When considering subunit vaccines, the most commonly utilized are recombinant proteins. While theoretically simple, production is often protein-dependent, with different hosts sometimes required to produce different proteins [4]. DNA vaccines, however, are simple plasmids, and their preparation is identical regardless of the protein they encode. DNA is also very stable, and can be stored long-term at ambient temperature, something that may not be possible for other types of vaccines.

Where has the field moved to in the 21 years since 1990? Certainly, there have been major achievements. The first "real-world" demonstration of the efficacy of DNA vaccination was a stunning success. This was a vaccine against the West Nile Virus, which had been spreading throughout the United States since 1999, killing many birds [5]. As the virus moved across the country, there were grave concerns for the endangered California condors, of which approximately 200 remained at the time. Under direction from the US Center for Disease Control and Prevention, Aldevron manufactured an experimental DNA vaccine expressing West Nile virus pre-membrane/membrane and envelope proteins, and the condors were immunized intramuscularly. The condors showed excellent neutralizing antibodies 60 days post-vaccination which continued to increase in titer until approximately one year post-vaccination and, and the birds were fully protected from West Nile viral infection [6].

However, many other experimental procedures failed to provide immunity. We have previously summarized these issues in developing vaccines against parasites [7], but similar issues can be generalized for all pathogen vaccines. One particular issue was the seemingly lower immunogenicity in larger mammals than in mice, which is presumably in part due to simple relative dose differences. However, the recent registration of DNA vaccines for treating melanoma in dogs (ONCEPT™) and West Nile Virus in horses (West Nile-Innovator®) demonstrate that it is not only the size of the host that influences efficacy. While expression levels may be somewhat important (for example, in cases where codon bias decreases translation), strong immunity is probably more reliant on the numbers of host cells that are transfected. Here we review some of the methods used to enhance the immunogenicity of DNA vaccines. Figure 1 shows the points in the immune induction process where intervention should have a positive effect on the magnitude of immune responses.

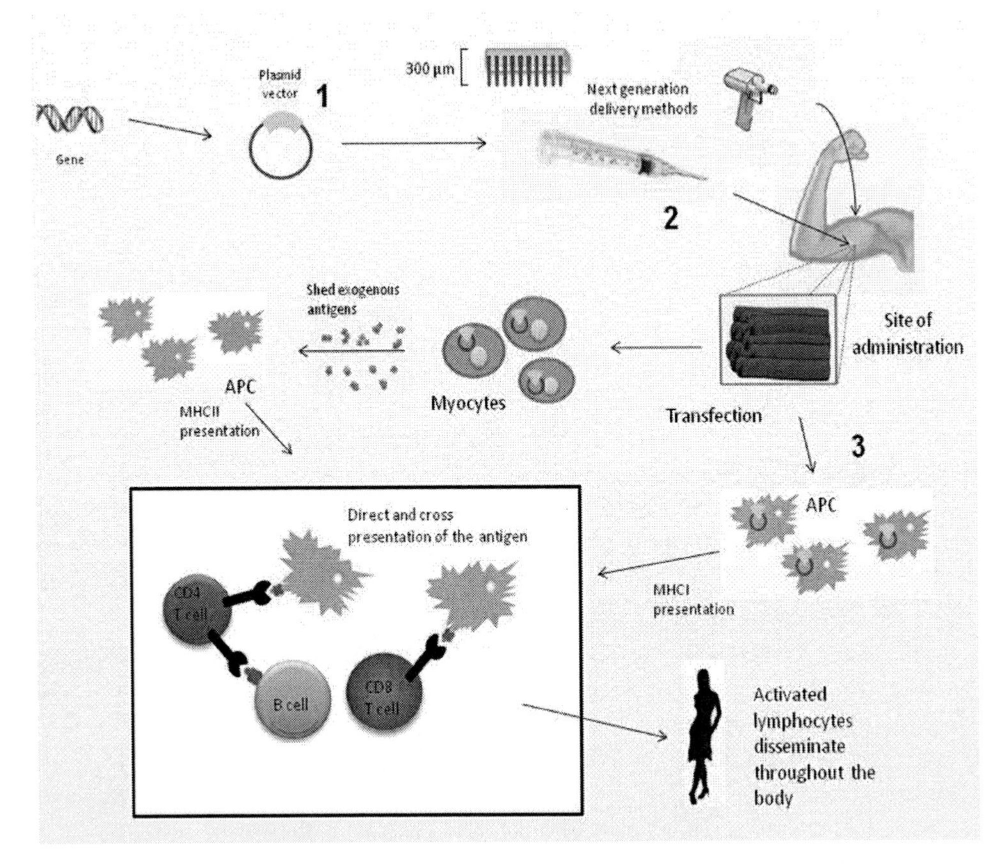


Figure 1. Schematic representation of the induction of immune responses after DNA vaccination. Sites of intervention to increase immunity include: 1. Increase the expression of encoded protein, 2. Use delivery methods that maximize plasmid delivery, and 3. Improve the transfection frequency of APC's, particularly dendritic cells. Modified from [8].

# Improving Immunogenicity of Plasmid DNA Vaccines

The route by which DNA vaccines are delivered to the body will ultimately determine the immune response elicited. The objective of a vaccine is to induce both cellular and humoral immune responses via a simple delivery method. The most widely used routes of delivery are intramuscular (IM) and intradermal (ID) injection, however in the case of plasmid DNA vaccines (henceforth termed simply DNA vaccines), these simple routes of administration have proven insufficient to elicit a potent immune response.

In the past, DNA vaccines have fallen short on the ability to induce both humoral and cellular immune responses to an appropriate level. Transfection efficiency has been one of the barriers, as the levels of transferred DNA via IM injection have been relatively low when compared to the amount administered [9-10]. Improving immunogenicity allows for smaller doses of DNA vaccines to be used which influences safety and costs of the vaccines – a feature of interest when the vaccines are to be used in developing countries. It will also overcome some of the issues of relative dose in large mammals. We will look at DNA routes of delivery and review how the latest delivery systems together with co-stimulatory molecules are being exploited as potent immune stimulators.

## Routes of Delivery of Plasmid DNA Vaccines

*Intramuscular injection* uses a needle and syringe to deliver DNA vaccines directly into the muscle, usually the deltoid/upper arm. After IM needle injection, monocytes and dendritic cells (DCs) are transfected and the immune system is activated via the MHC I and/or MHC II pathways followed by priming of $CD8^+$ cytotoxic T-cells (CTLs) and $CD4^+$ T-helper cells. In a study comparing IM needle injection with delivery via tattooing of a HPV type 16 DNA vaccine, it was tattooing that elicited a higher cellular immune response when adjuvants were used [11]. The enhanced immune responses of tattooing can be attributed to a greater number of cells being transfected, with the delivery of vaccine covering a larger surface area of the skin, and the involvement of inflammatory responses resulting from trauma caused to the skin during tattooing. Although this method of delivery may not be widely accepted for use in humans, it is a feasible method for prophylaxis in the veterinary industry.

*Intradermal injection* is into the dermal layer of the skin, which sits beneath the top layer of the skin (epidermis), but above the subcutaneous later. Cells residing here include fibroblasts, mast cells, macrophages, infiltrating T-lymphocytes and dermal DCs (dDCs). This layer also contains blood and lymph vessels. There are two subsets of dDCs – $CD14^+$ and $CD1a^+$ which act as antigen-capturing cells and alert the immune system of incoming danger signals and express MHC class II molecules [12-13]. Although ID injection has been shown to induce stronger immune responses with a lower antigen dose compared to IM [14], skilled personnel are required to administer the vaccine via this method. Thus, new methods of ID delivery such as microinjection have been developed to overcome the need for trained healthcare professionals.

*Mucosal* vaccines may be delivered to the oral, nasal, lung, rectal, inner ear, conjunctival and vaginal mucosa. This route of delivery is suited to diseases that are transmitted and initiated at the mucosal surface. Mucosal vaccine delivery may induce both systemic and mucosal immunity and induces $CD8^+$ CTLs, $CD4^+$ T-helper cells and the production of secretory IgA (s-IgA), which is otherwise not induced by parenteral delivery routes. When a vaccine enters the mucosal space they are transported through microfold (M) cells within the mucosal-associated lymphoid tissue (MALT), and then can be taken up by resident macrophages and dendritic cells and presented to T-cells which provide help for the production of s-IgA from B-cells [15]. The mucosal surface is being investigated as a delivery route for diseases such as HIV [16] and Herpes simplex virus [17], which are both transmitted via this route.

In many cases, these delivery routes do not induce the desired immune response to an encoded antigen. Physical delivery systems have thus been developed for use in conjunction with these methods to increase transfection efficiency as well as enhance cellular and humoral immune responses to DNA vaccines. Of course there are limitations to these systems and the ongoing difficulty of designing DNA vaccines against organisms with high sequence variations and high mutation rates, such as Hepatitis C and HIV, also has an influence on the success of a vaccine. Physical delivery methods and the use of co-stimulatory molecules to enhance vaccine efficacy will be discussed.

## Physical Delivery Systems of Plasmid DNA Vaccines

In order to increase transfection efficiency and subsequent immune response after vaccination, physical delivery methods have been used together with established routes of delivery. These delivery systems include needle-free methods as well as a broad range of nanoparticle delivery systems which will be discussed below.

## Needle-Free Delivery Methods

*Jet injection* (JI) is a needle-free delivery system that delivers liquid droplets or solid particles via a high pressure narrow stream penetrating into the skin. An advantage of JI delivery of liquid is the dispersion of the fluid into the tissue allowing it to make contact with a larger surface area and hence with more immune cells, instead of remaining in a fluid-filled sphere after needle-syringe delivery [15]. This may be the mechanism behind the enhanced immune responses after JI delivery compared to delivery by needle-syringe. The advantages of JI surround the speed and ease of use for delivery of DNA vaccines. This becomes especially important during times of pandemic where a large population would require vaccination in a short period of time. For similar reasons, JI may be favorable over needle-syringe delivery for vaccination of farm animals, with the use of JI for delivery of DNA vaccines to cattle shown previously [18]. Another benefit is safety of JI with a diminished risk of transmission of infectious diseases from needle-stick injuries [15]. JI typically elicits immune responses comparable to or better than that of needle-syringe delivery. A study comparing three IM needle-free delivery methods: Biojector® 2000, Mini-Ject™, and needle-syringe, using a DNA vaccine against HIV cynomolgus monkeys, showed comparable immune responses across all three methods. These results suggested that JI did not enhance immunogenicity compared with needle-syringe delivery, however the advantages of JI delivery such as ease of use, avoidance of needle-stick injury favor the use of JI over IM injection [19]. JI has successfully been used for the delivery of veterinary DNA vaccines and induced IgG1 and IgG2a antibodies against West Nile Virus [20], and high neutralizing antibody titers against Japanese Encephalitis Virus (JEV) [21]. A DNA vaccine against rabies delivered ID by JI induced longer lasting neutralizing antibodies than a commercial vaccine delivered by either subcutaneous or IM injection [22]. An Ebola virus DNA vaccine delivered IM by JI induced antibodies and $CD4^+$ and $CD8^+$ T-cells in a phase I clinical trial [23]. JI has also been assessed as a delivery method for vaginal vaccines using a needle-free device currently used for daily administration of insulin in humans [24]. This study showed that JI delivery induced INF-γ and vaginal IgA responses in both mice and rabbits as well as an increase in gene expression compared to needle-syringe delivery. JI allowed for a lower amount of DNA to be administered to elicit an appropriate immune response [20, 22]. In one case it was shown that increasing the amount of DNA vaccine administered did not improve the immune response [22]. Although such small doses can be delivered by electroporation (EP) and gene gun (both are discussed below), JI is favored as it is a quicker, easier, more convenient and less invasive than other devices.

*A gene gun* is a particle delivery device used for biolistic transfection of DNA-coated particles into cells and was originally used for plant transfection. Gene gun delivery of DNA vaccines in humans and animals involves DNA-coated particles (usually gold), and a device requiring a trained operator. These two factors alone make the system challenging to introduce in developing countries. Gene gun delivers DNA into the epidermal layer of the skin where it may transfect both professional (Langerhans cells), and non-professional (keratinocytes) antigen presenting cells (APCs) [25]. Gene gun delivery has provided complete protection in small animal [26], and non-human primate animal models [25], but it has also been shown to be inferior to EP delivery [27]. Th2 responses dominate with gene gun delivery [28-30] which may not be protective, while co-delivery of DNA vaccines with various adjuvants to shift the Th2-biased response still does not stimulate a Th1 response [29, 31]. Another concern with this system is the pain associated with biolistic delivery as well as the inability of the body to degrade and/or excrete the gold particles. However, studies have shown that pain may be minimized with the use of a low pressure gene gun, and the use of biodegradable particles such as chitosan, may be an effective replacement for gold particles [29, 32]. Although the gene gun is still being trialed for DNA vaccine delivery, it has been shown that the epidermis may not be optimal for all target antigens as DNA reaches only the keratinocytes and Langerhans cells which may in fact prevent the induction of $CD8^+$ T cells [33].

Electropermeabilization or *electroporation* (EP) has been widely used as the standard method of transfecting DNA into bacterial cells *in vitro* for the past 25 years. More recently however, EP has been used in humans for the transfer of genetic material, namely therapeutics and DNA into tissue and organs. The principle of EP vaccine delivery is based on an electric field pulse delivered usually concomitantly or after IM injection of the DNA vaccine. Akin to the reaction of bacterial cells *in vitro*, an electric field pulse applied *in vivo* results in transient destabilization of the cell membrane (lasting as long as the duration of the pulse), allowing DNA to enter the cell and make its way to the nucleus [34]. The electric pulse is delivered to the muscle via an electronic device such as the TriGrid™ Delivery System (TDS) [27, 35-36], or the Easy Vax™ DNA vaccine delivery system [37]. EP is able to delivery both small molecules, thought to be transferred via diffusion, and larger molecules such as DNA, which enter the cell in a slightly more complex fashion. It is hypothesized that the entry of large molecules such as DNA occurs via a multistep process. The first stage involves the interaction of the DNA with the plasma membrane, followed by transfer across the membrane [38]. The precise mechanism of entry, however, is still under debate [39]. EP has been used in conjunction with IM delivery of DNA vaccines to increase transfection efficiency which is much lower when IM [40] or ID [41] delivery is used alone. An increase in transfection efficiency leads to an increased uptake of DNA into the cells and a subsequent increase in the level of protein expression.

When delivered using IM/EP a two-vector DNA vaccine against HIV type 1 had a 2.5-$\log_{10}$ increase in antibody response and a significant decrease in DNA persistence. There was also a 50-to 200-fold increase in potency after DNA was delivered by IM/EP compared to IM delivery alone [42]. The ADVAX vaccine against HIV-1 also showed an improvement in $CD4^+$ and $CD8^+$ T-cell responses when the vaccine was delivered by IM/EP [36]. Cell-mediated immunity and humoral immune responses against Human Papilloma virus (HPV) [27], tuberculosis [43], malaria [44], bovine diarrhea virus [45], and anthrax [46] have also been enhanced with the use of EP. EP has been shown to be safe for the delivery of DNA vaccines in large animals such as cattle [35] and sheep [47], non-human primates [48], and small animals such as mice [37]. There are currently clinical trials evaluating the use of EP for delivery of DNA vaccines in humans against H5N1 avian influenza virus (ClinicalTrials.gov ID: NCT01142362) and malaria (ClinicalTrials.gov ID: NCT01169077).

Although the vast majority of studies indicate the safety of DNA vaccines delivered by EP [49], there is evidence of integration of the DNA into the host genome following EP, however the integration even was of low frequency, less than 1000/220,000 copies of the DNA were integrated [10]. The electric field pulse delivered during EP may cause damage to the cells (seen as histological changes), however there is evidence that the damage EP causes to cells may in fact serve as an adjuvant by attracting APCs and other cells of the immune system to the site [47]. Any safety concerns related to the biodistribution of the DNA and the persistence of plasmid have been ruled out in numerous studies [36, 42, 50].

# Expression Optimization

*Codon optimization* is the replacement of rarely-available codons within the wild-type DNA (or RNA) sequence with codons used more frequently by the intended host, whilst retaining the same amino acid sequence. Codon optimization has been used to increase the level of protein expression in DNA vaccines against the Mycobacterial antigen Ag85B [51], malaria [52], HIV [53] and H1N1 swine influenza [54]. Although protein expression is generally enhanced by codon optimization, the extent to which protective immunity is enhanced varies and may be antigen-specific and also dependent on the algorithm used to select the codons to be replaced.

The *Kozak sequence* (GCC) GCC A/G CC AUG G, first described in 1987 [55], serves to enhance translation efficiency by regulating ribosomal scanning to ensure the AUG start codon is recognized. This sequence is generally conserved within most vertebrates and has been included upstream of the initiation codon in plasmid DNA vaccines against poxvirus [56] and HIV-1 subtype B [57], and it has been successful in improving gene translation of DNA vaccines in horses [58]. Kozak sequences are not widely used in plasmid DNA vaccines, and there is also controversy over the sequence being "consensus" across various species [59]. However, like many plasmid DNA vaccine components, the effects of its inclusion to improve initiation of gene translation may be antigen-and/or host-dependent.

*CpG motifs* are known to activate innate, humoral and cellular immune responses and have been shown to induce a Th1-biased immune responses in draining lymph nodes when included in a DNA vaccine that would otherwise induce a Th2-biased response when delivered by gene gun [60]. Immunization with CpG motifs has been shown to activate the immune system via Toll-like receptor-9, stimulate B cells and Natural Killer cells as well as induce $CD8^+$ T-cell responses [61-62]. CpG has successfully enhanced immune responses of vaccines against *Toxoplasma gondii* in sheep [63], viral hemorrhagic septicaemia virus in fish [64] and Porcine reproductive and respiratory syndrome virus [65].

*Signal peptides* can be used to direct an antigen to a desired location within the cell (cytosol or endoplasmic reticulum (ER)) or to the cell membrane or extracellular environment (via the ER). Depending on where the antigen is targeted will alter how the antigen is processed by APCs and thus result in a different immune response [66]. The use of a secretory signal derived from tissue plasminogen activator (TPA) which secreted a JEV antigen into the culture medium in transfection studies, has been shown to significantly enhance immune responses compared to the non-secretory form [67]. The cytokine responses were also altered with re-direction of the antigen to produce both a Th1 and Th2 response, where the non-secretory form could only induce a Th1 response. Another study using TPA to secrete a protein from Bovine viral diarrhea virus had similar success [68]. The use of a signal peptide to deliver a hepatitis C antigen resulted in similar immune responses to the non-secreted DNA vaccine indicating that the presence of a signal peptide will not enhance immunogenicity in all DNA vaccines [69]. There is however, clear evidence to support the use of a secretory signal to enhance the immunogenicity of DNA vaccines such as in cases where increasing DNA vaccine dose may not be feasible for the host species [68].

*Antigen targeting* involves coupling an antigen to an antibody with affinity to a specific subset of APCs, in order to deliver the antigen directly to the DCs of interest. The antibody recognizes specific cell surface receptors on DCs and its fusion to the antigen results in the shuttling of the antigen to the DC where it can be presented [70]. In some cases the antigen may be chemically conjugated to the antibody fragment [71], or expressed as a fusion protein [72]. An HIV gag antigen has been fused to a single-chain Fv antibody fragment targeting the DEC205 antigen uptake receptor of DCs in mice [72]. Vaccination with this DNA vaccine increased numbers of INF-$\gamma^+$ cells producing $CD4^+$ and $CD8^+$ T-cells, and induced a 10-fold increase in antibody levels. In a study using CTLA-4 as the targeting molecule to DCs, significantly increased immune responses were observed [73]. Despite its success in animal models, antigen targeting has yet to be tested in humans. A consideration that must be made is the possibility of inducing T-cell tolerance with the use of monoclonal antibody targeting. However, a way to circumvent tolerance is to include maturation signals such as anti-CD40 monoclonal antibody [70-71].

## Cytokines

Immune responses to DNA vaccines can be enhanced by co-delivery of co-stimulatory molecules and/or cytokines. Generally, these molecules are plasmid encoded and co-expressed along with the antigen, but they may be encoded on a separate plasmid and transfected concomitantly with the DNA vaccine. Interleukin-12 (IL-12) has been shown to increase INF-$\gamma^+$ cells required for immunity against viruses and intracellular bacteria [74-75], and IL-18 has been shown to stimulate T-cell proliferation in a chicken model [76]. Bias toward a Th1 response can be facilitated through co-delivery of IL-18 [76] and GM-CSF [77], whilst RANTES, an inflammatory chemokine, has been shown to stimulate a mixed Th1/Th2 response in mice [78]. Co-delivery of cytokines such as IL-12 in combination with delivery via non-conventional methods such as *in vivo* EP [74] and gene gun [75], have been shown to further improve immune responses by inducing IFN-γ-dominated immune responses. Another study has shown that co-delivery of a cytokine from a bicistronic DNA vaccine, showed an enhanced immune response than co-delivery of the same cytokine from a monocistronic vector [76]. Cytokines such as IL-23, which induces proliferation of memory T-cells, have been shown to facilitate long-term protective responses in a mouse model [78], whilst other cytokines have lead to a decrease in antibody responses as seen in the use of IL-10, involved in immune suppression [77]. In some instances, the presence of cytokines has not enhanced these responses at all, as demonstrated with the use of complement fragment C3d [31]. Taken together, these studies show that immune responses may not only be altered and/or enhanced by the cytokine co-delivered, but also how the cytokine is delivered and the method used to administer the plasmid(s) has an influence.

Adjuvants such as the 4-1BBL and OX40L [79] of the tumor necrosis factor receptor family have been shown to enhance antigen-specific cell-mediated responses, whilst M11L, an anti-apoptotic molecule was able to promote cell survival and persistent antigen expression in a HIV vaccine [80]. However, the anti-apoptotic gene Bcl-XL reduced vaccine efficacy in a malarial DNA vaccine [81], indicating that adjuvants that enhance vaccine efficacy in one system, such as for viral vaccines, may not be a suitable candidate for use in non-viral vaccines.

# Delivery Vectors for DNA Vaccines

Vaccine vectors can be in many different manifestations from a whole organism such as a modified virus or bacterium to a gene sequence or a plasmid. Each vector has its own benefits and limitations which dictates its usefulness in the health industry. The strength of such vectors is that in cases where the naked DNA vaccine does not elicit systemic protective responses against re-infection, the use of a vector system can target and stimulate the immune system into a specific and localized response. There are three main types of vectors – bacterial, viral, plasmid DNA and RNA (reviewed in [82]). One of the main advantages of using live bacterial and viral vectors to deliver heterologous antigens is that they have been used successfully as autonomous vaccines and their efficacy has already been demonstrated.

## Bacterial Vectors

Live bacterial vectors are an ideal delivery system for DNA vaccines as they are easy to manufacture, there is no need for purification and the vector itself acts as a natural adjuvant. The rationale for using such vectors was elucidated in the 1980s and subsequent research investigated the suitability of natural invasive intestinal pathogens as vectors such as *Salmonella*, *Shigella* and *Listeria* species (reviewed in [83]). There have been various levels of success using these vectors, and a potential limitation has been demonstrated that the success of heterologous delivery is dependent on the presence of both eukaryotic and cryptic prokaryotic promoters in the DNA vaccine, which may be advantageous in further applications [84].

Of the bacterial vectors, *Salmonella typhimurium* is one of the most promising with many studies using it to deliver a wide range of heterologous antigens from various organisms. Protective responses against Avian Reovirus (ARV) have been elicited using *S. typhimurium* as a delivery vector. When 6 day old chickens were vaccinated twice using the capsid protein σC DNA vaccine in the Salmonella vector, a significantly high level of humoral responses was detected (compared to the control vaccinated chickens) and upon ARV challenge, 66.7% of the chickens were protected from the disease [85]. This type of delivery is not limited to viral pathogens but can be used for eukaryotic parasites. One recent report used a *Salmonella* vector to deliver an antigen from *Trichinella spiralis,* a nematode parasite that is acquired from the ingestion of larvae in contaminated undercooked meats. This prophylactic vaccine demonstrated a significant reduction to 29.8% in adult worm burden and 34.2% muscle larvae following challenge with *T. spiralis* larvae, compared with mice that were immunized with the empty *Salmonella* vector [86]. This DNA vaccine expressed an immunodominant antigen, Ts87 that is excreted from adult worms in the gut and in the study the vaccine elicited a localized IgA response along with a balanced Th1/Th2 immune response that gave partial protection against challenge [86].

Therapeutic applications have also been investigated using live bacterial vectors. Using *Salmonella typhimurium* as a delivery vector, Bai *et al.,* (2010) constructed a fusion-expression plasmid where two fragments of the hormone somatostatin (growth hormone inhibiting hormone) were fused with fragments of the hepatitis B surface antigen and co-stimulatory granulocyte/macrophage colony-stimulating factor (GM-CSF). The *Salmonella*–vectored delivery resulted in systemic IgG immune responses and promoted growth in female mice. It also increased first litter size, milk production and growth rate of offspring in female mice [87]. The use of stimulatory molecules such as GM-CSF with the hepatitis B antigen demonstrates that this system is successful as both a prophylactic application with the stimulation of immune responses and therapeutic application with the increase in litter size.

It is widely accepted that low-copy number plasmids are the most stable for the delivery of DNA vaccines in live bacterial vectors [88-89]. However one recent study described using the Operator-Repressor Titration (ORT) plasmid maintenance system in *Salmonella typhimurium* to stably deliver a high-copy number plasmid expressing the *Mycobacterium tuberculosis* gene mpt64 in mice [90]. The delivery of this antigen was successful to the mucosal surfaces and induced significantly higher levels of antigen-specific IFN-γ from the bacterial vector than that of intramuscular delivery of plasmid DNA. Protective responses against tuberculosis challenge were demonstrated from this bacterial delivery of the antigen [90].

Although there has been a lot of focus on the natural invasive intestinal pathogens, a few recent studies have reported on other bacterial species that may be able to improve delivery of these DNA vaccines. *In vitro* studies have been carried out using *Lactococcus lactis* and *Streptococcus gordonii,* non-invasive bacteria, to deliver fluorescent protein as a DNA vaccine to mammalian cells [91]. Tao *et a*l., (2010) found that uptake of red fluorescent plasmid carrying bacteria by colon cancer (Caco-2) cells was significantly improved by using cell wall-disrupting agents such as glycine, penicillin, lysozyme or electroporation alone or in varying combinations on the *L. lactis* or *S. gordonii* vectors. This study suggests that there is potential in non-invasive bacterial vectors which should be explored in tandem with the natural invasive intestinal bacteria.

## Viral Vectors

Many viral vectors have been examined as delivery systems for heterologous genes. The two most common are adenovirus and poxvirus. Vaccinia virus, a poxvirus, was used successfully in the global eradication of smallpox [92] and adenoviruses are very commonly used for gene delivery and immunization purposes [93]. However because of this widespread usage there is a high percentage of the world population that have prior-immunological memory of these viruses through either vaccination or natural infection. The prevalence of pre-existing immunity to the most common viruses has been reported in several studies [94-96]. This limitation has been linked to the failure of the Merck STEP phase II test of concept trial for their Adenovirus Type 5 (Ad5) based HIV vaccine which did not induce protective responses against HIV infection or reduce viral loads in infected patients [95]. A recent study confirmed that the international prevalence of pre-existing immunity to different adenovirus types can severely compromise its efficacy as the majority (85.2%) of the 1904 participants were sero-positive for neutralizing antibodies against Ad5 [97].

Choosing a virus from a different animal reservoir avoids this limitation of pre-existing immunity as the subjects are immunologically naive to the vaccine vector. An example of this is the Vaccinia Ankara virus (rMVA) which is a popular choice for investigation for use in species other than humans, such as cattle and horses. One study used rMVA expressing the immediate early gene focused on eliciting protection against equine herpesvirus-1 (EHV-1). Although it did not result in full protection, reduction of clinical symptoms and cell associated viremia were observed indicating that if this vaccine is coupled with a structural EHV-1 gene such as gB and gC full protection may be achieved [98]. Another study using rMVA as a vector investigated the effect of pre-existing immunity on elicited immune responses. They reported that although the pre-existing immunity decreased SIV-specific $CD8^+$ and $CD4^+$ T cell responses, the SIV-specific humoral immunity remained constant. In addition, pre-existing immunity did not diminish the partial control of an SIV challenge mediated by the DNA/MVA vaccine induced immunity [96]. In another study DNA priming with MVA boost-vaccinations demonstrated a high magnitude of virus specific $CD4^+$ T cell responses and priming with vector-associated proteins increased clonal expansion of HIV Env-specific $CD4^+$ T cells suggesting that these pre-existing effects can be manipulated by choosing an appropriate agent for the booster vaccination [99].

As the problem of pre-existing immunity appears to be surmountable by selection of suitable boosters recombinant adenovirus vectors are still being developed for several severe diseases such as hemorrhagic fever caused by the Ebola virus. A trial was performed using the Ebola viral glycoprotein as a heterologous antigen to provide protection against the 5 different species of Ebola virus found in Africa. The replication defective adenovirus vector elicited GP-specific $CD4^+$ and $CD8^+$ responses that could be detected 4 weeks post vaccination for high and low doses of the vaccine. This study also showed that the pre-existing immunity to adenovirus can affect the production of specific responses to the heterologous antigens [94]. For a potential malaria vaccine one study used both simian adenoviral and poxviral vectors to deliver multiple DNA constructs of merozoite surface protein 1 (MSP-1) from *Plasmodium falciparum*. In preclinical trials, these vectored vaccines generated potent cellular $CD8^+$ T cell immune responses and high titer antibodies specific to MSP-1 in mice which demonstrated growth inhibitory properties [100].

## DNA and RNA Replicons

Over the last twenty years, bacterial plasmid DNA, which is circular, extrachromosomal genetic material, has been used for genetic vaccination. However, it was found that non-structural viral genes, when combined in linear form with a heterologous gene (rather than the original structural viral genes), would produce multiple copies of the sequence without the need of an origin of replication or selective markers [101].

The replicons, either DNA or RNA, are based on any encoded heterologous antigen and the viral proteins required to replicate the construct *in vivo* which have been sourced from numerous viruses such as alphaviruses (Figure 2). These are the latest generation of genetic-based vaccines.

An early study established that a DNA based subgenomic replicon from porcine reproductive and respiratory syndrome virus (PRRSV) could generate RNA replicons using the PRRSV expression genes present in the DNA replicon. It could also express heterologous genes and when *in vivo* could elicit responses specific to both the reporter gene and part of the replicon resulting in significantly higher IFN-γ than conventional DNA vaccines [29]. The early vector systems based on the replicons of alpha viruses necessitated the production and manipulation of recombinant RNA *in vitro* and these RNA constructs required stabilization due to their propensity for degradation *in vivo* [102-103]. Modifications were made to eliminate these highly technical and tedious procedures and DNA replicons were developed which retained its high level of heterologous protein expression and strong elicited immune responses [104].

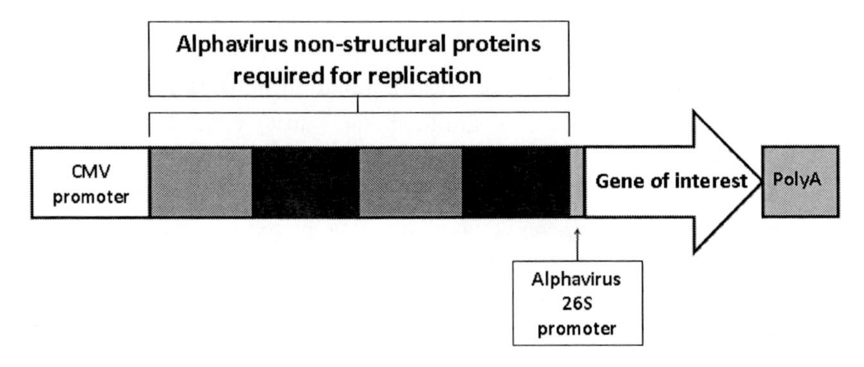


Figure 2. Schematic representation of the components of a replicon. There are three main components: 26S promoter, genes encoding the non-structural viral proteins required for genetic replication and the gene of interest.

Alpha virus DNA replicons have become a promising vector system and have shown great efficacy when used in a prime/boost schedule with a heterologous boost. The use of an alpha virus replicon expressing E2 glycoprotein from Classical Swine Fever virus (CSFV) with an adenovirus boost expressing the same antigen provided full protection against challenge in pigs. The vaccinated pigs had high titers of CSFV-specific neutralizing antibodies and comparable $CD4^+$ and $CD8^+$ T cell proliferation and at challenge showed no clinical signs of illness [105]. An alpha virus replicon expressing the Gn glycoprotein of Rift Valley fever virus was used in mice to determine the efficacy of this system. There was correlation between the high neutralizing titers induced and the reduction in clinical signs of illness for use of the replicon alone and in a prime/boost schedule with a Gn-C3d DNA construct as the prime vaccination.

The Gn replicon vaccine was able to induce robust cellular responses that were Gn-specific and protective whereas the traditional DNA plasmid vaccine alone was unable to [106].

A benefit of these constructs is that these alpha virus replicon vectors can be used in conjunction with existing vaccines to improve their efficacy and usability. One alpha virus DNA replicon was used to deliver goat pox structural protein. When used as a DNA priming vaccine, it greatly reduced the adverse reactions normally associated with the current live vaccine AV4, which was administered as the booster vaccination. This vaccination schedule also provided partial protection at challenge [104]. Co-delivery of GM-CSF and administration with an aluminum phosphate adjuvant significantly increased specific humoral responses and survival rates for a *Clostridium botulinum* neurotoxin DNA replicon vaccine [107].

Minimalistic, immunogenetically defined gene expression (MIDGE) vectors are another recent development in DNA vaccination. MIDGE vectors are linear, double stranded DNA molecules which have ends which are covalently closed with a single stranded hairpin loop at both ends. They have three main features; a CMV immediate-early promoter, the gene of interest and a polyA site. For MIDGE vectors designed to induce a Th1 biased immune response, vectors contain a peptide localization sequence (NLS) that facilitates translocation to the nucleus, bound to one of the hairpin loops (Figure 3) [108]. This type of vaccine has been reported to be dose dependent for humoral responses and elicited favorable humoral and cell mediated responses from intradermal administration rather than traditional intramuscular delivery especially when covalently bound with SAINT-18 which is a strong transfection reagent [109]. MIDGE vectors have been demonstrated to be equally effective at transfection as traditional plasmid DNA both *in vitro* [110] and *in vivo* [109]. The linear nature of this delivery system lends itself to covalent attachment of co-stimulatory molecules [111].

The use of biological vectors to deliver heterologous antigens has been reported with varying levels of success, however they continue to be developed to increase stability and immunogenicity with many going on to clinical trials. These vectors can simulate a natural infection which often leads to broad and strong humoral and cellular responses. The difficulty of prior exposure to these vaccine vectors can be completely circumvented by using synthetic materials to induce specific immune responses. This is the promising area of using nanoparticles to transfer genetic material.

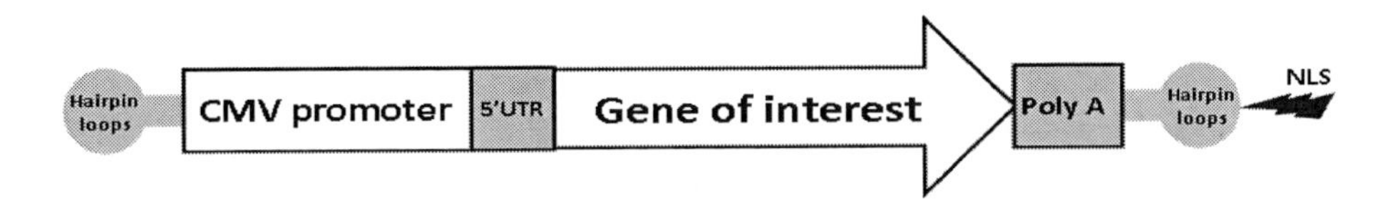


Figure 3. Schematic representation of minimalistic immunogenetically defined gene expression (MIDGE) vectors. These are linear vectors with hairpin loops at the ends which can have covalently bonded co-stimulatory molecules such as the nuclear localization sequence (NLS) to increase transfection efficiency *in vivo*. Adapted from [108].

## Nano-Sized Vehicles for Gene Delivery

As mentioned above, one of the major hindrances to generating immune responses by a DNA vaccine is the limited ability to transfect a high proportion of cells, in other words, the level of gene-transfer [112]. The inefficient delivery of DNA molecules results in low levels of antigenic protein expression, limiting the potency of the vaccine [113]. Gene delivery must also be directed to the correct cell types and pathways to generate strong immunity. Although viral vectors are known to be highly efficient in transfection, they have some limitations due to the possible risk of gene integration into the host chromosome [112] and the lack of specificity in targeting cells [114].

Several methodologies have been exploited to increase the immunogenicity of a DNA vaccine as mentioned in earlier sections, but one of the most rapidly growing fields of study is the use of nanotechnology. With the technology advancing in recent years, micro-or nano-sized materials are gaining interest as potential gene carriers for new DNA vaccine delivery systems. Micro-or nano-sized particles can be made from various materials including inorganic and biodegradable polymers, often coming in the form of spheres or capsules. The encapsulation of DNA using polymer based particles, or complexation of DNA by adsorption onto the cationic material has been found to significantly enhance gene transfer since they can provide stability to DNA and promote improved cellular uptake of the particulate-DNA complex. This section will explore these properties. Given that this is a relatively new technology, some time will be spent on describing the various materials used for DNA vaccine delivery.

Encapsulation of DNA is an effective strategy as it can provide protection from extracellular and intracellular nuclease degradation. DNA can also be released at a controlled rate by designing a varying particle degradation rate for different sets of particles, therefore providing prolonged or repeated release of DNA and thus gene availability over a longer period [115]. Similarly, adsorption of DNA to the surface of a positively charged cationic particle made from lipid molecules and polymers has also shown to enhance transfection and gene delivery, due to the electrostatic interaction of particles with cell membranes [116-117]. The ability of particulates to protect DNA is not limited to *in vivo* degradation, but also offers stability of DNA during the particle formulation, which offers an attractive advantage to the development of vaccines. Particles can also be designed to accommodate extra functionalities such as adjuvant properties and other immunostimulatory agents by conjugation.

Materials that have shown the potential to deliver DNA include inorganic materials such as calcium phosphate and silica [118-120], and biomaterials such as cationic lipids and polymers that can adsorb or encapsulate DNA by electrostatic interactions [116, 121-125]. Particularly in recent years, cationic polymers and lipids have gained most of the focus and have been extensively investigated due to their biodegradability and biocompatibility. Such nanoparticles synthesized with polymers and lipids are referred to as polyplexes and lipoplexes, respectively [126-128].

## Inorganic to Biomaterial – Past to Future

Early reports indicated that the method of calcium phosphate co-precipitation of DNA was a simple and effective method for improving *in vivo* transfection [118]. Calcium ions are known to form ionic complexes with phosphate in the DNA backbone, therefore it functions to stabilize DNA during gene delivery [129]. Calcium phosphate particles can be as small as 80 nm in diameter [118-119]. When used in a viral vaccine trial calcium phosphate nanoparticles have also shown potential as a vaccine adjuvant [130]. Other inorganic based materials also gained focus as they may be utilized as a gene carrier, and one of these was surface modified silica particles [120]. It was later discovered that compared to calcium phosphate, silica particles possessed a weaker ability to deliver DNA. The transfection efficiency however can be enhanced by formulating hybrid particles with other transfection agents [131] or chemical alterations for surface charge and amine content to specifically enhance uptake by cells, particularly DCs [132].

Gold nanoparticles have also been extensively studied, primarily in the context of "gene gun" delivery, as described in earlier sections. In recent years, gold nanoparticles with different modifications have been assessed. Examples of such modifications include surface modification with the conjugation of cationic polymers such as polyethylenimine (PEI) or poly(ethylene glycol) (PEG) to increase the efficiency of cellular uptake [133], or the controlled plasmid DNA release from the nanoparticles by an exposure to laser irradiation [134]. It is thought that these kinds of nanoparticles can enhance immunogenicity over prolonged periods of time since they may remain in the tissue at the site of injection [135].

Increasing attention has been directed towards the synthesis of particles consisting of multiple materials possessing different properties to further increase the DNA payload and enhance cellular uptake, over that achieved by one material alone. Several hybrids of inorganic-organic or organic-biomaterials have been tested, but an example of an interesting formulation of nanoparticles is the use of magnesium hydroxide and aluminum hydroxide to create layered double hydroxides (LDHs). LDH is a uniquely layered material consisting of positively charged cationic layers with charge-balancing anions between layers [114]. This material is already in use for drug delivery studies to assess compatibility with diverse (negatively charged) drug molecules and controlled discharge of loaded drug into an acidic environment [136]. They also possess a number of advantageous properties, such as fast cellular uptake and nuclear localization, and interestingly, targeting of specific intracellular organelles depending on modifications of the particle morphology [137], which makes these inorganic hybrid nanoparticles attractive drug delivery vehicles. Li and colleagues have made an observation of the uptake of LDHs by monocyte-derived DCs within 3 h, with no adverse effects on the cells [138]. Furthermore, aluminum compounds in the LDH nanoparticles are well known vaccine adjuvants (Alum) and may significantly enhance the potency of DNA vaccines [139]. Aluminum hydroxide as an adjuvant can also play a critical role in the maturation of DCs. Upon the uptake of LDH nanoparticles by endocytosis, the immature DCs expressed surface markers indicating maturation. One team has observed an increase in the expression of CD86 and CD40 (co-stimulatory markers) [138], while in another study a significant increase in the expression of CD80 (co-stimulatory marker) was observed while there was no effect on CD40 in [140]. Aluminum hydroxide can also have an impact on macrophages as it can increase the expression level of CD83 (a maturation marker) and CD86 [141].

There are number of other materials available as potential gene carrier candidates in addition to the inorganic compounds mentioned above. Many groups have shifted their focus to the use of cationic micro/nanoparticles using biodegradable polymers and lipids. The cationic particles posses a positively charged surface and have the property of adsorbing anionic molecules such as DNA and negatively charged proteins [116-117, 142]. Compared to naked DNA, both types of cationic particles have been shown to result in a higher transfection efficiency when applied to non-transformed human epithelial cells [143] and again they have shown the same effect upon intramuscular injection, resulting in increased potency of the DNA vaccine [121, 144-145]. The cationic moiety of the particles allows DNA to condense thus providing protection from the physiological environment. The positively charged particle-DNA complex then mediates the transfection process by effectively binding to the negatively charged cell membrane, allowing the complex to enter the cytoplasm via various intracellular pathways [124].

Cationic lipids have shown their potential as a gene delivery vehicle since the initial published report by Felgner in 1987 [122]. Cationic lipid molecules can be structured into various shapes such as micelles (spheres), bilayers (sheets) or liposomes. Liposomes, which are sometimes termed virus-like particles, are large unilamellar or multi-lamellar vesicles with a phospholipid bilayer membrane and an aqueous core that is composed of non-viral lipids [123, 135, 146]. Liposomes are often constructed with phospholipids along with various lipid molecules. DNA is encapsulated into liposomes with its cationic charge facilitating the process, along with the presence of another lipid molecule such as dioleylphosphatidylethanolamine (DOPE) or cholesterol. DOPE and cholesterol can increase the rigidity of the lipid bilayer complex, and finally the addition of hydrocarbon chains can offer better efficiency for both *in vitro* and *in vivo* transfection [121]. Lipsomes first demonstrated their potential as a vaccine carrier in 1997, in the immunization study using mice vaccinated with liposome encapsulating plasmids encoding a hepatitis B antigen. The application of liposomes afforded improved humoral and cellular immunity [147]. Their application is extensive in clinical trials for gene therapy, and they are currently the primary method used as a non-viral delivery vector (data from June 2010) [54]. There are some cautions in this technology, however. Many cationic lipids are considered toxic-they can interact with serum proteins inhibiting the transfection process, activate the complement system, and have difficulties to disperse well in the target tissue because of their large particle size [121, 148]. There is a need for continued development of improved materials for safer liposomes.

In recent years, polymers have gained interest as an emerging material for gene delivery, rather than lipids. Examples of polymers studied and utilized for this purpose include; poly(L-lysine) (PLL) [125], polyethylenimine (PEI) [124], polybutyl cyanoacrylate (PBCA) [116] and also a naturally synthesized polymer, chitosan [117, 149-150]. Polymers are one of the most attractive materials for gene delivery, due to their biodegradability, biocompatibility and the ease of formulation. They can electrostatically bind and complex plasmid DNA to form polyplexes, and can facilitate increased humoral and cellular immune responses due to enhanced DNA uptake by DCs [125]. As the plasmid DNA is relatively large, the formulation of polymers with terminal primary amines are preferred to facilitate binding of DNA [151]. The potential of these polymers as a particulate gene carrier was first observed in 2000, using poly(lactide-co-glycolide) (PLG) microparticles formulated along with the cationic surfactant hexadecyltrimethylammonium bromide (CTAB), which delivered DNA into APCs [144]. Microparticles often range from 1 – 10 μm diameter therefore they can be efficiently taken up by APCs via phagocytosis [152]. They are generally phagocytozed by macrophages and DCs since they cannot promote endocytosis due to their relatively large size [153].

It is important to note that the microencapsulation can provide control over the particles degradation rate, influencing the DNA release. The rate of release of encapsulated DNA can be significantly changed depending on the polymer formulation, perhaps with the addition of protein encapsulation [154-155]. A variety of poly(lactic-co-glycolic) acid (PLGA) copolymers are commercially available which differ in their lactic/glycolic monomer compositions [156]. Selecting the appropriate compositions, the degradation rate of the particles can be extended from weeks to months [154]. The controlled release rate may be very important as different timing can elicit differing immune responses [157]. After the microparticle is taken up by DCs, gene expression of encapsulated plasmid DNA will initiate. However, it requires at least one or two days for DCs to migrate to the draining lymph nodes, hence early expression and presentation of encoded protein may result in immune tolerance [157].

Although cationic molecules are used widely in the area of gene delivery and drug delivery, there are some concerns that the high positive charge of the particle may be cytotoxic [158] hence the amount delivered should be kept to a minimum. However, this may result in a less potent vaccine. One possible strategy to circumvent this cytotoxicity is to add charge-reducing agents to the nanoparticle. Introduction of poly(ethylene glycol) (PEG) moieties to cationic nanoparticles is known to decrease the surface charge, while its potency as a DNA vaccine is maintained, and in fact it was demonstrated to increase antigen expression [143, 159].

## Size Determines the Fate of Particles

In all vaccination studies, the targeting of specific cells can significantly enhance the magnitude of the immune response [157, 160]. DCs are the most effective primary APCs and they can control the subsequent immune response by the internalization and presentation of antigens through the MHC class I and class II pathways to $CD4^+$ and $CD8^+$ T lymphocytes. Therefore, designing an antigen delivery system to increase transfection of DCs will be beneficial [161].

Several strategies using micro and nanoparticles have been applied to target APCs effectively, however it appears that nanoparticles may have some advantages over microparticles. It is known that nanoparticles can generally provide greater transfection efficiency than microparticles [162], and this is probably due to the particle size influencing the route of entry to the cells [163]. Nanoparticles have demonstrated their effectiveness in uptake across skin and mucosal membrane to target the mucosal associated lymphatic tissues (MALT) [164]. The positive charge of a cationic particle surface can readily interact with negatively charged glycoproteins on the cell surface enabling non-specific cellular uptake by APCs [159, 165], with uptake facilitated by the natural function of DCs to constantly sample the environment [166]. Polymeric nanoparticles for gene delivery usually range from 10 to

100 nm in diameter, and can be synthesized as either spheres with polymeric matrices, or capsules with a polymeric shell and an inner core [165]. These easily fit within the range of 500 nm or less, which is the optimal size uptake by APC [167-168]. For nanoparticles, several routes of entry are available, mediated either via phagocytosis, macropinocytosis, or receptor-mediated endocytosis [165-166, 169]. Furthermore, it was demonstrated that the nanoparticle-plasmid DNA complex within a very tightly controlled range of around 40 nm, close to viral size, could enhance *in vitro* DC transfection rate, resulting in the induction of enhanced immunogenicity [125]. These nanoparticles were able to induce high levels of $CD8^+$ T lymphocytes and antibody production against the plasmid encoded antigen, compared with the use of bacterial size microparticles (>1000 nm).

Secondly, the fate of nanoparticle trafficking to the draining lymph nodes is thought to be size-dependent [170-172]. Nanoparticles that are 200 nm or less can freely drain to the lymph nodes, where they can be taken up by the local DCs and initiate antigen presentation to T lymphocytes [172]. Manolova and colleagues (2008), found that fluorescent-tagged nanoparticles injected intracutaneously were rapidly trafficked to the draining lymph nodes within 2 hours of injection, whereas microparticles ranging from 200 to 2 μm entered the subcapsular sinus 8 hours after the injection. Reflecting the time required for DCs to migrate, it is thought that DCs in the injection site were responsible for the microparticle transport [172]. These findings show the potential of small nanoparticles as a suitable carrier for gene delivery to the resident DCs in draining lymph nodes, and this DC-independent trafficking can decrease the time to antigen presentation [157]. Small nanoparticles can also be conjugated with protein antigens and immunostimulatory substances such as synthetic CpG oligonucleotides to induce stronger cellular immune responses [170, 173].

Lastly, synthesis of nano-sized particles gives the potential to manipulate the type of immune response to be elicited [174]. The adjuvant activities of nanoparticles have been observed [135, 175], and the formulation with polymers has been shown to give a significant adjuvant effect in the activation of APCs upon transfection. Polymers such as PLGA and PEI in the microparticle system have also shown to be very effective in adjuvant activity [176]. Lecithin-based nanoparticles prepared with the addition of glyceryl monostearate (GSM) were used in a protein antigen delivery system, and the adjuvanticity of these nanoparticles was found to be much stronger than that of a traditionally used adjuvant, Alum (aluminum hydroxide) [177]. The adjuvant effect was not only limited to the material formulating the nanoparticles, but the smaller size was found to possess more potent adjuvant activity than that of larger microparticles [125, 144]. The particle size not only has an influence on the transfection and trafficking of desired cell types, but also influences the Th1 and Th2 cytokine responses [174]. Ovalbumin-conjugated nanobeads size ranging from 40 to 49 nm were compared to those from 93 to 101 nm, and the maximal Th1 priming was found with the smaller size range, while Th2 and consequent IL-4 induction were favored by the larger size range [174]. These findings strongly suggest that the size of particles is having a significant effect on the immunological outcome. Smaller nanoparticles ranging around 40 nm can induce a strong cellular response, whereas larger microparticles greater than 1 μm can promote phagocytosis and elicit humoral responses and antibody production [168, 178]. The consideration of particle size can significantly impact on vaccine design, and exploiting these avenues to attain the desired immune responses can lead to the development of a potent DNA vaccine carrier.

## Conclusion

DNA vaccines have come a long way in 21 years, and have entered the commercial sphere with the availability of several vaccines. However, there is as yet no DNA vaccine available for use in humans. This is not for want of trying, and surely the day will come when such vaccines are available. The issue of relatively poor immunogenicity that we have addressed in this review is being overcome by some judicious vaccine design, utilizing a variety of accessory molecules or adjuvants such as CpG motifs. The real breakthroughs however are likely to come from nanomedicine, whereby vaccines can be designed in such a way that the immunological fate is controlled. This is where the next 21 years will take us.

## References

[1] Wolff, J.A., et al., Direct gene transfer into mouse muscle in vivo. *Science,* 1990. 247(4949 Pt 1): p. 1465-8.

[2] Tang, D.C., M. De Vit, and S.A. Johnston, Genetic immunization is a simple method for eliciting an immune response. *Nature,* 1992. 356(6365): p. 152-4.

[3] Ulmer, J.B., et al., Heterologous protection against influenza by injection of DNA encoding a viral protein. *Science,* 1993. 259(5102): p. 1745-9.

[4] Jayaraj, R. and P. Smooker, So you Need a Protein -A Guide to the Production of Recombinant Proteins. *3, 28-34. The Open Veterinary Science Journal*, 2009. 2: p. 28-34.

[5] Turell, M.J., et al., DNA vaccine for West Nile virus infection in fish crows (Corvus ossifragus*). Emerg. Infect. Dis.*, 2003. 9(9): p. 1077-81.

[6] Chang, G.J., et al., Prospective immunization of the endangered California condors (Gymnogyps californianus) protects this species from lethal West Nile virus infection. *Vaccine,* 2007. 25(12): p. 2325-30.

[7] Smooker, P.M., et al., DNA vaccines and their application against parasites--promise, limitations and potential solutions. *Biotechnol. Annu. Rev.*, 2004. 10: p. 189-236.

[8] Baird FJ, et al., *DNA Vaccines: a modern-day vaccine revolution.* Immunogenicity, in press, 2011.

[9] Huygen, K., Plasmid DNA vaccination. *Microbes Infect.,* 2005. 7(5-6): p. 932-8.

[10] Wang, Z., et al., Detection of integration of plasmid DNA into host genomic DNA following intramuscular injection and electroporation. *Gene Ther.,* 2004. 11(8): p. 711-21.

[11] Pokorna, D., I. Rubio, and M. Muller, DNA-vaccination via tattooing induces stronger humoral and cellular immune responses than intramuscular delivery supported by molecular adjuvants. *Genet. Vaccines Ther.*, 2008. 6: p. 4.

[12] Valladeau, J. and S. Saeland, Cutaneous dendritic cells. *Seminars in Immunology*, 2005. 17(4): p. 273-283.

[13] Bal, S.M., et al., Advances in transcutaneous vaccine delivery: Do all ways lead to Rome? *Journal of Controlled Release*, 2010. 148(3): p. 266-282.

[14] Van Damme, P., et al., Safety and efficacy of a novel microneedle device for dose sparing intradermal influenza vaccination in healthy adults. *Vaccine*, 2009. 27(3): p. 454-459.

[15] Giudice, E.L. and J.D. Campbell, Needle-free vaccine delivery. *Advanced Drug Delivery Reviews*, 2006. 58(1): p. 68-89.

[16] Belyakov, I.M. and J.D. Ahlers, Mucosal Immunity and HIV-1 Infection: Applications for Mucosal AIDS Vaccine Development. *Curr. Top Microbiol. Immunol.,* 2011.

[17] Mott, K.R., S.L. Wechsler, and H. Ghiasi, Ocular infection of mice with an avirulent recombinant HSV-1 expressing IL-4 and an attenuated HSV-1 strain generates virulent recombinants in vivo. *Mol. Vis*., 2010. 16: p. 2153-62.

[18] Carter, E.W. and D.E. Kerr, Optimization of DNA-based vaccination in cows using green fluorescent protein and protein A as a prelude to immunization against staphylococcal mastitis. *J. Dairy Sci*., 2003. 86(4): p. 1177-86.

[19] Rao, S.S., et al., Comparative evaluation of three different intramuscular delivery methods for DNA immunization in a nonhuman primate animal model. *Vaccine,* 2006. 24(3): p. 367-373.

[20] Ishikawa, T., et al., Co-immunization with West Nile DNA and inactivated vaccines provides synergistic increases in their immunogenicities in mice. *Microbes Infect.,* 2007. 9(9): p. 1089-95.

[21] Imoto, J.-i., et al., Needle-free jet injection of small doses of Japanese encephalitis DNA and inactivated vaccine mixture induces neutralizing antibodies in miniature pigs and protects against fetal death and mummification in pregnant sows. *Vaccine*, 2010. 28(46): p. 7373-7380.

[22] Bahloul, C., et al., Field trials of a very potent rabies DNA vaccine which induced long lasting virus neutralizing antibodies and protection in dogs in experimental conditions. Vaccine, 2006. 24(8): p. 1063-1072.

[23] Martin, J.E., et al., A DNA vaccine for Ebola virus is safe and immunogenic in a phase I clinical trial. *Clin. Vaccine Immunol.,* 2006. 13(11): p. 1267-77.

[24] Kanazawa, T., et al., Effects of menstrual cycle on gene transfection through mouse vagina for DNA vaccine. *Int. J. Pharm*., 2008. 360(1-2): p. 164-70.

[25] Loudon, P.T., et al., *GM-CSF* increases mucosal and systemic immunogenicity of an H1N1 influenza DNA vaccine administered into the epidermis of non-human primates. *PLoS One,* 2010. 5(6): p. e11021.

[26] Hu, J., et al., Protective immunity with an E1 multivalent epitope DNA vaccine against cottontail rabbit papillomavirus (CRPV) infection in an HLA-A2.1 transgenic rabbit model. *Vaccine,* 2008. 26(6): p. 809-16.

[27] Best, S.R., et al., Administration of HPV DNA vaccine via electroporation elicits the strongest CD8+ T cell immune responses compared to intramuscular injection and intradermal gene gun delivery. *Vaccine,* 2009. 27(40): p. 5450-5459.

[28] Wang, S., et al., The relative immunogenicity of DNA vaccines delivered by the intramuscular needle injection, electroporation and gene gun methods. *Vaccine,* 2008. 26(17): p. 2100-10.

[29] Huang, H.N., et al., Transdermal immunization with low-pressure-gene-gun mediated chitosan-based DNA vaccines against Japanese encephalitis virus. *Biomaterials,* 2009. 30(30): p. 6017-25.

[30] Williman, J., et al., The use of Th1 cytokines, IL-12 and IL-23, to modulate the immune response raised to a DNA vaccine delivered by gene gun. *Vaccine*, 2006. 24(21): p. 4471-4.

[31] Weiss, R., et al., Differential effects of C3d on the immunogenicity of gene gun vaccines encoding Plasmodium falciparum and Plasmodium berghei MSP1(42). *Vaccine*, 2010. 28(28): p. 4515-22.

[32] Lee, P.W., et al., The use of biodegradable polymeric nanoparticles in combination with a low-pressure gene gun for transdermal DNA delivery. *Biomaterials*, 2008. 29(6): p. 742-51.

[33] Gaffal, E., et al., Comparative evaluation of CD8+CTL responses following gene gun immunization targeting the skin with intracutaneous injection of antigen-transduced dendritic cells. *Eur. J. Cell. Biol.*, 2007. 86(11-12): p. 817-26.

[34] Rols, M.P., Mechanism by which electroporation mediates DNA migration and entry into cells and targeted tissues. *Methods Mol. Biol.,* 2008. 423: p. 19-33.

[35] van Drunen Littel-van den Hurk, S., et al., Electroporation-based DNA transfer enhances gene expression and immune responses to DNA vaccines in cattle. *Vaccine,* 2008. 26(43): p. 5503-5509.

[36] Dolter, K.E., et al., Immunogenicity, safety, biodistribution and persistence of ADVAX, a prophylactic DNA vaccine for HIV-1, delivered by in vivo electroporation. *Vaccine*. In Press, Uncorrected Proof.

[37] Hooper, J.W., et al., Smallpox DNA vaccine delivered by novel skin electroporation device protects mice against intranasal poxvirus challenge. *Vaccine,* 2007. 25(10): p. 1814-1823.

[38] Bodles-Brakhop, A.M., R. Heller, and R. Draghia-Akli, Electroporation for the delivery of DNA-based vaccines and immunotherapeutics: current clinical developments. *Mol. Ther.*, 2009. 17(4): p. 585-92.

[39] Escoffre, J.M., et al., What is (still not) known of the mechanism by which electroporation mediates gene transfer and expression in cells and tissues. *Mol. Biotechnol.*, 2009. 41(3): p. 286-95.

[40] Ongkudon, C.M., J. Ho, and M.K. Danquah, Mitigating the looming vaccine crisis: production and delivery of plasmid-based vaccines. *Crit. Rev. Biotechnol.*, 2010.

[41] Pavselj, N. and V. Preat, DNA electrotransfer into the skin using a combination of one high-and one low-voltage pulse. *J. Control Release*, 2005. 106(3): p. 407-15.

[42] Luckay, A., et al., Effect of plasmid DNA vaccine design and in vivo electroporation on the resulting vaccine-specific immune responses in rhesus macaques. *J. Virol.,* 2007. 81(10): p. 5257-69.

[43] Li, Z., et al., DNA electroporation prime and protein boost strategy enhances humoral immunity of tuberculosis DNA vaccines in mice and non-human primates. *Vaccine,* 2006. 24(21): p. 4565-4568.

[44] Dobaño, C., et al., Enhancement of antibody and cellular immune responses to malaria DNA vaccines by in vivo electroporation. *Vaccine*, 2007. 25(36): p. 6635-6645.

[45] van Drunen Littel-van den Hurk, S., et al., Electroporation enhances immune responses and protection induced by a bovine viral diarrhea virus DNA vaccine in newborn calves with maternal antibodies. *Vaccine,* 2010. 28(39): p. 6445-6454.

[46] Luxembourg, A., et al., Potentiation of an anthrax DNA vaccine with electroporation. *Vaccine,* 2008. 26(40): p. 5216-5222.

[47] Scheerlinck, J.P.Y., et al., In vivo electroporation improves immune responses to DNA vaccination in sheep. *Vaccine,* 2004. 22(13-14): p. 1820-1825.

[48] Livingston, B.D., et al., Comparative performance of a licensed anthrax vaccine versus electroporation based delivery of a PA encoding DNA vaccine in rhesus macaques. *Vaccine*, 2010. 28(4): p. 1056-1061.

[49] Bråve, A., et al., Biodistribution, persistence and lack of integration of a multigene HIV vaccine delivered by needle-free intradermal injection and electroporation. *Vaccine,* 2010. 28(51): p. 8203-8209.

[50] Roos, A.K., et al., Skin electroporation: effects on transgene expression, DNA persistence and local tissue environment. *PLoS One,* 2009. 4(9): p. e7226.

[51] Ko, H.J., et al., Optimization of codon usage enhances the immunogenicity of a DNA vaccine encoding mycobacterial antigen Ag85B. *Infect. Immun*., 2005. 73(9): p. 5666-74.

[52] Dobaño, C., et al., Plasmodium: Mammalian codon optimization of malaria plasmid DNA vaccines enhances antibody responses but not T cell responses nor protective immunity. *Experimental Parasitology*, 2009. 122(2): p. 112-123.

[53] Megati, S., et al., Modifying the HIV-1 env gp160 gene to improve pDNA vaccine-elicited cell-mediated immune responses. *Vaccine,* 2008. 26(40): p. 5083-5094.

[54] Tenbusch, M., et al., Codon-optimization of the hemagglutinin gene from the novel swine origin H1N1 influenza virus has differential effects on CD4+ T-cell responses and immune effector mechanisms following DNA electroporation in mice. *Vaccine,* 2010. 28(19): p. 3273-3277.

[55] Kozak, M., At least six nucleotides preceding the AUG initiator codon enhance translation in mammalian cells. *J. Mol. Biol.,* 1987. 196(4): p. 947-50.

[56] Moise, L., et al., VennVax, a DNA-prime, peptide-boost multi-T-cell epitope poxvirus vaccine, induces protective immunity against vaccinia infection by T cell response alone. *Vaccine*, 2011. 29(3): p. 501-511.

[57] Yan, J., et al., Enhanced cellular immune responses elicited by an engineered HIV-1 subtype B consensus-based envelope DNA vaccine. *Mol. Ther.,* 2007. 15(2): p. 411-21.

[58] Olafsdottir, G., et al., In vitro analysis of expression vectors for DNA vaccination of horses: the effect of a Kozak sequence. *Acta Vet Scand.,* 2008. 50: p. 44.

[59] Nakagawa, S., et al., Diversity of preferred nucleotide sequences around the translation initiation codon in eukaryote genomes. Nucleic Acids Res., 2008. 36(3): p. 861-71.

[60] Liu, L., et al., CpG motif acts as a 'danger signal' and provides a T helper type 1-biased microenvironment for DNA vaccination. *Immunology*, 2005. 115(2): p. 223-30.

[61] Coban, C., et al., Effect of plasmid backbone modification by different human CpG motifs on the immunogenicity of DNA vaccine vectors. *J. Leukoc. Biol.,* 2005. 78(3): p. 647-55.

[62] Berchtold, C., et al., Superior protective immunity against murine listeriosis by combined vaccination with CpG DNA and recombinant Salmonella enterica serovar typhimurium. *Infect. Immun*., 2009. 77(12): p. 5501-8.

[63] Li, B., et al., Immunological response of sheep to injections of plasmids encoding Toxoplasma gondii SAG1 and ROP1 genes. *Parasite Immunol.,* 2010. 32(9-10): p. 671-83.

[64] Martinez-Alonso, S., et al., The introduction of multi-copy CpG motifs into an antiviral DNA vaccine strongly up-regulates its immunogenicity in fish. *Vaccine*, 2010.

[65] Quan, Z., et al., Plasmid containing CpG oligodeoxynucleotides can augment the immune responses of pigs immunized with porcine reproductive and respiratory syndrome killed virus vaccine. *Vet. Immunol. Immunopathol.,* 2010. 136(3-4): p. 257-64.

[66] Rice, J., et al., Manipulation of pathogen-derived genes to influence antigen presentation via DNA vaccines. *Vaccine*, 1999. 17(23-24): p. 3030-8.

[67] Ashok, M.S. and P.N. Rangarajan, Protective efficacy of a plasmid DNA encoding Japanese encephalitis virus envelope protein fused to tissue plasminogen activator signal sequences: studies in a murine intracerebral virus challenge model. *Vaccine,* 2002. 20(11-12): p. 1563-70.

[68] Liang, R., et al., Immunization with plasmid DNA encoding a truncated, secreted form of the bovine viral diarrhea virus E2 protein elicits strong humoral and cellular immune responses. *Vaccine*, 2005. 23(45): p. 5252-62.

[69] Inchauspe, G., et al., Plasmid DNA expressing a secreted or a nonsecreted form of hepatitis C virus nucleocapsid: comparative studies of antibody and T-helper responses following genetic immunization. *DNA Cell Biol.,* 1997. 16(2): p. 185-95.

[70] Caminschi, I., M.H. Lahoud, and K. Shortman, Enhancing immune responses by targeting antigen to DC. *Eur. J. Immunol.,* 2009. 39(4): p. 931-8.

[71] Romani, N., et al., Targeting of antigens to skin dendritic cells: possibilities to enhance vaccine efficacy. *Immunol. Cell Biol.,* 2010. 88(4): p. 424-30.

[72] Nchinda, G., et al., The efficacy of DNA vaccination is enhanced in mice by targeting the encoded protein to dendritic cells. *J. Clin. Invest.,* 2008. 118(4): p. 1427-36.

[73] Boyle, J.S., J.L. Brady, and A.M. Lew, Enhanced responses to a DNA vaccine encoding a fusion antigen that is directed to sites of immune induction. *Nature*, 1998. 392(6674): p. 408-11.

[74] Hirao, L.A., et al., Combined effects of IL-12 and electroporation enhances the potency of DNA vaccination in macaques. *Vaccine*, 2008. 26(25): p. 3112-20.

[75] Yoshida, S., et al., DNA vaccine using hemagglutinating virus of Japan-liposome encapsulating combination encoding mycobacterial heat shock protein 65 and interleukin-12 confers protection against Mycobacterium tuberculosis by T cell activation. *Vaccine*, 2006. 24(8): p. 1191-204.

[76] Chen, H.Y., et al., IL-18-mediated enhancement of the protective effect of an infectious laryngotracheitis virus glycoprotein B plasmid DNA vaccine in chickens. *J. Med. Microbiol.,* 2010.

[77] Yen, H.H. and J.P. Scheerlinck, Co-delivery of plasmid-encoded cytokines modulates the immune response to a DNA vaccine delivered by in vivo electroporation. *Vaccine,* 2007. 25(14): p. 2575-82.

[78] Williman, J., et al., DNA fusion vaccines incorporating IL-23 or RANTES for use in immunization against influenza. *Vaccine*, 2008. 26(40): p. 5153-8.

[79] Xiao, C., et al., Enhanced protective efficacy and reduced viral load of foot-and-mouth disease DNA vaccine with co-stimulatory molecules as the molecular adjuvants. *Antiviral Research*, 2007. 76(1): p. 11-20.

[80] Su, J., et al., Inclusion of the viral anti-apoptotic molecule M11L in DNA vaccine vectors enhances HIV Env-specific T cell-mediated immunity. *Virology,* 2008. 375(1): p. 48-58.

[81] Bergmann-Leitner, E.S., et al., *Molecular adjuvants for malaria DNA vaccines based on the modulation of host-cell apoptosis. Vaccine,* 2009. 27(41): p. 5700-5708.
[82] Liu, M.A., Immunologic basis of vaccine vectors. *Immunity,* 2010. 33(4): p. 504-15.
[83] Becker, P.D., M. Noerder, and C.A. Guzman, *Genetic immunization: bacteria as DNA vaccine delivery vehicles. Hum .Vaccin,* 2008. 4(3): p. 189-202.
[84] Gahan, M.E., et al., Bacterial antigen expression is an important component in inducing an immune response to orally administered Salmonella-delivered DNA vaccines. *PLoS One*, 2009. 4(6): p. e6062.
[85] Wan, J., et al., Immunogenicity of a DNA vaccine of Avian Reovirus orally delivered by attenuated Salmonella typhimurium. *Res. Vet. Sci.,* 2010.
[86] Yang, Y., et al., Oral vaccination with Ts87 DNA vaccine delivered by attenuated Salmonella typhimurium elicits a protective immune response against Trichinella spiralis larval challenge. *Vaccine,* 2010. 28(15): p. 2735-2742.
[87] Bai, L.-Y., et al., Effects of immunization against a DNA vaccine encoding somatostatin gene (pGM-CSF/SS) by attenuated Salmonella typhimurium on growth, reproduction and lactation in female mice. *Theriogenology,* 2010. In Press, Corrected Proof.
[88] Galen, J.E., et al., A new generation of stable, nonantibiotic, low-copy-number plasmids improves immune responses to foreign antigens in Salmonella enterica serovar Typhi live vectors. *Infect. Immun.,* 2010. 78(1): p. 337-47.
[89] Loessner, H., et al., Improving live attenuated bacterial carriers for vaccination and therapy. *International Journal of Medical Microbiology*, 2008. 298(1-2): p. 21-26.
[90] Huang, J.-M., et al., Oral delivery of a DNA vaccine against tuberculosis using operator-repressor titration in a Salmonella enterica vector. *Vaccine,* 2010. 28(47): p. 7523-7528.
[91] Tao, L., et al., A novel plasmid for delivering genes into mammalian cells with noninvasive food and commensal lactic acid bacteria. *Plasmid,* 2010. In Press, Corrected Proof.
[92] Behbehani, A.M., The smallpox story: life and death of an old disease. *Microbiol. Rev.,* 1983. 47(4): p. 455-509.
[93] Hartman, Z.C., D.M. Appledorn, and A. Amalfitano, Adenovirus vector induced innate immune responses: Impact upon efficacy and toxicity in gene therapy and vaccine applications. *Virus Research*, 2008. 132(1-2): p. 1-14.
[94] Ledgerwood, J.E., et al., A replication defective recombinant Ad5 vaccine expressing Ebola virus GP is safe and immunogenic in healthy adults. *Vaccine*, 2010. 29(2): p. 304-313.
[95] Buchbinder, S.P., et al., Efficacy assessment of a cell-mediated immunity HIV-1 vaccine (the Step Study): a double-blind, randomised, placebo-controlled, test-of-concept trial. *The Lancet*, 2008. 372(9653): p. 1881-1893.
[96] Kannanganat, S., et al., Preexisting Vaccinia Virus Immunity Decreases SIV-Specific Cellular Immunity but Does Not Diminish Humoral Immunity and Efficacy of a DNA/MVA Vaccine. *J. Immunol.,* 2010.
[97] Mast, T.C., et al., International epidemiology of human pre-existing adenovirus (Ad) type-5, type-6, type-26 and type-36 neutralizing antibodies: Correlates of high Ad5 titers and implications for potential HIV vaccine trials. *Vaccine,* 2010. 28(4): p. 950-957.

[98] Soboll, G., et al., Vaccination of ponies with the IE gene of EHV-1 in a recombinant modified live vaccinia vector protects against clinical and virological disease. *Veterinary Immunology and Immunopathology*, 2010. 135(1-2): p. 108-117.

[99] Sun, Y., et al., Recombinant vector-induced HIV/SIV-specific CD4+ T lymphocyte responses in rhesus monkeys. *Virology,* 2010. 406: p. 48-55.

[100] Goodman, A.L., et al., New candidate vaccines against blood-stage Plasmodium falciparum malaria: prime-boost immunization regimens incorporating human and simian adenoviral vectors and poxviral vectors expressing an optimized antigen based on merozoite surface protein 1. *Infect. Immun.,* 2010. 78(11): p. 4601-12.

[101] Strauss, J.H. and E.G. Strauss, The alphaviruses: gene expression, replication, and evolution. *Microbiol. Rev.,* 1994. 58(3): p. 491-562.

[102] Kuhn, A.N., et al., Phosphorothioate cap analogs increase stability and translational efficiency of RNA vaccines in immature dendritic cells and induce superior immune responses in vivo. *Gene Ther.,* 2010. 17(8): p. 961-71.

[103] Thornburg, N.J., et al., Vaccination with Venezuelan equine encephalitis replicons encoding cowpox virus structural proteins protects mice from intranasal cowpox virus challenge. *Virology,* 2007. 362(2): p. 441-52.

[104] Zheng, M., et al., Immunogenicity and protective efficacy of Semliki forest virus replicon-based DNA vaccines encoding goatpox virus structural proteins. *Virology*, 2009. 391(1): p. 33-43.

[105] Sun, Y., et al., Enhanced immunity against classical swine fever in pigs induced by prime-boost immunization using an alphavirus replicon-vectored DNA vaccine and a recombinant adenovirus. *Veterinary Immunology and Immunopathology*, 2010. 137(1-2): p. 20-27.

[106] Bhardwaj, N., M.T. Heise, and T.M. Ross, Vaccination with DNA plasmids expressing Gn coupled to C3d or alphavirus replicons expressing gn protects mice against Rift Valley fever virus. *PLoS Negl Trop Dis*, 2010. 4(6): p. e725.

[107] Li, N., et al., Enhancement of the immunogenicity of DNA replicon vaccine of Clostridium botulinum neurotoxin serotype A by GM-CSF gene adjuvant. Immunopharmacol. *Immunotoxicol.*, 2011. 33(1): p. 211-9.

[108] Lopez-Fuertes, L., et al., DNA vaccination with linear minimalistic (MIDGE) vectors confers protection against Leishmania major infection in mice. *Vaccine,* 2002. 21(3-4): p. 247-57.

[109] Endmann, A., et al., Immune response induced by a linear DNA vector: Influence of dose, formulation and route of injection. *Vaccine,* 2010. 28(21): p. 3642-3649.

[110] Moreno, S., et al., DNA immunisation with minimalistic expression constructs. *Vaccine*, 2004. 22(13-14): p. 1709-1716.

[111] Schirmbeck, R., et al., Priming of immune responses to hepatitis B surface antigen with minimal DNA expression constructs modified with a nuclear localization signal peptide. *J. Mol. Med.,* 2001. 79(5-6): p. 343-50.

[112] Putnam, D., Polymers for gene delivery across length scales. *Nat. Mater.,* 2006. 5(6): p. 439-451.

[113] Ulmer, J.B., B. Wahren, and M.A. Liu, Gene-based vaccines: recent technical and clinical advances. *Trends in Molecular Medicine,* 2006. 12(5): p. 216-222.

[114] Li, A., et al., The use of layered double hydroxides as DNA vaccine delivery vector for enhancement of anti-melanoma immune response. *Biomaterials*, 2011. 32(2): p. 469-477.
[115] Wang, D., et al., Encapsulation of plasmid DNA in biodegradable poly(,-lactic-co-glycolic acid) microspheres as a novel approach for immunogene delivery. *Journal of Controlled Release*, 1999. 57(1): p. 9-18.
[116] Duan, J., et al., Cationic polybutyl cyanoacrylate nanoparticles for DNA delivery. *J. Biomed.Biotechnol.*, 2009. 2009: p. 149254.
[117] Jayakumar, R., et al., Chitosan conjugated DNA nanoparticles in gene therapy. *Carbohydrate Polymers*, 2010. 79(1): p. 1-8.
[118] Roy, I., et al., Calcium phosphate nanoparticles as novel non-viral vectors for targeted gene delivery. *International Journal of Pharmaceutics*, 2003. 250(1): p. 25-33.
[119] Welzel, T., W. Meyer-Zaika, and M. Epple, Continuous preparation of functionalised calcium phosphate nanoparticles with adjustable crystallinity. *Chemical Communications*, 2004(10): p. 1204-1205.
[120] Kneuer, C., et al., Silica nanoparticles modified with aminosilanes as carriers for plasmid DNA. *International Journal of Pharmaceutics,* 2000. 196(2): p. 257-261.
[121] Tranchant, I., et al., Physicochemical optimisation of plasmid delivery by cationic lipids. *The Journal of Gene Medicine*, 2004. 6(S1): p. S24-S35.
[122] Felgner, P.L., et al., Lipofection: a highly efficient, lipid-mediated DNA-transfection procedure. *Proceedings of the National Academy of Sciences of the United States of America,* 1987. 84(21): p. 7413-7417.
[123] Gregoriadis, G., et al., Vaccine Entrapment in Liposomes. *Methods,* 1999. 19(1): p. 156-162.
[124] Lungwitz, U., et al., Polyethylenimine-based non-viral gene delivery systems. *European Journal of Pharmaceutics and Biopharmaceutics*, 2005. 60(2): p. 247-266.
[125] Minigo, G., et al., Poly-l-lysine-coated nanoparticles: A potent delivery system to enhance DNA vaccine efficacy. *Vaccine,* 2007. 25(7): p. 1316-1327.
[126] Henriques, A.M., et al., Effect of cationic liposomes/DNA charge ratio on gene expression and antibody response of a candidate DNA vaccine against Maedi Visna virus. *International Journal of Pharmaceutics*, 2009. 377(1-2): p. 92-98.
[127] Vachutinsky, Y. and K. Kataoka, PEG-based Polyplex Design for Gene and Nucleotide Delivery. *Israel Journal of Chemistry,* 2010. 50(2): p. 175-184.
[128] Huang, Y., et al., A recoding method to improve the humoral immune response to an HIV DNA vaccine. *PLoS One,* 2008. 3(9): p. e3214.
[129] Truong-Le, V.L., et al., Gene Transfer by DNA-Gelatin Nanospheres. *Archives of Biochemistry and Biophysics*, 1999. 361(1): p. 47-56.
[130] He, Q., et al., Calcium phosphate nanoparticle adjuvant. *Clin. Diagn. Lab. Immunol.*, 2000. 7(6): p. 899-903.
[131] Luo, D., et al., A self-assembled, modular DNA delivery system mediated by silica nanoparticles. *Journal of Controlled Release*, 2004. 95(2): p. 333-341.
[132] Gemeinhart, R.A., D. Luo, and W.M. Saltzman, Cellular Fate of a Modular DNA Delivery System Mediated by Silica Nanoparticles. *Biotechnology Progress*, 2005. 21(2): p. 532-537.
[133] Thomas, M. and A.M. Klibanov, Conjugation to gold nanoparticles enhances polyethylenimine's transfer of plasmid DNA into mammalian cells. *Proceedings of the*

*National Academy of Sciences of the United States of America*, 2003. 100(16): p. 9138-9143.

[134] Pissuwan, D., T. Niidome, and M.B. Cortie, The forthcoming applications of gold nanoparticles in drug and gene delivery systems. *Journal of Controlled Release*, 2011. 149(1): p. 65-71.

[135] Peek, L.J., C.R. Middaugh, and C. Berkland, Nanotechnology in vaccine delivery. *Advanced Drug Delivery Reviews*, 2008. 60(8): p. 915-928.

[136] Choy, J.-H., et al., Layered double hydroxide as an efficient drug reservoir for folate derivatives. *Biomaterials*, 2004. 25(15): p. 3059-3064.

[137] Xu, Z.P., et al., Subcellular compartment targeting of layered double hydroxide nanoparticles. *Journal of Controlled Release*, 2008. 130(1): p. 86-94.

[138] Li, A., et al., Signalling pathways involved in the activation of dendritic cells by layered double hydroxide nanoparticles. *Biomaterials*, 2010. 31(4): p. 748-56.

[139] Yu, Y.-Z., et al., Enhanced potency of individual and bivalent DNA replicon vaccines or conventional DNA vaccines by formulation with aluminum phosphate. *Biologicals,* 2010. 38(6): p. 658-663.

[140] Sokolovska, A., S.L. Hem, and H. HogenEsch, Activation of dendritic cells and induction of CD4+ T cell differentiation by aluminum-containing adjuvants. *Vaccine*, 2007. 25(23): p. 4575-4585.

[141] Rimaniol, A.-C., G. Gras, and P. Clayette, In vitro interactions between macrophages and aluminum-containing adjuvants. *Vaccine*, 2007. 25(37-38): p. 6784-6792.

[142] Kwon, Y.J., et al., Enhanced antigen presentation and immunostimulation of dendritic cells using acid-degradable cationic nanoparticles. *Journal of Controlled Release*, 2005. 105(3): p. 199-212.

[143] van den Berg, J.H., et al., Shielding the cationic charge of nanoparticle-formulated dermal DNA vaccines is essential for antigen expression and immunogenicity. *Journal of Controlled Release*, 2010. 141(2): p. 234-240.

[144] Singh, M., et al., Cationic microparticles: A potent delivery system for DNA vaccines. *Proceedings of the National Academy of Sciences of the United States of America,* 2000. 97(2): p. 811-816.

[145] O'Hagan, D., et al., Induction of potent immune responses by cationic microparticles with adsorbed human immunodeficiency virus DNA vaccines. *J. Virol.,* 2001. 75(19): p. 9037-43.

[146] Mui, B., L. Chow, and M.J. Hope, Extrusion Technique to Generate Liposomes of Defined Size, in *Methods in Enzymology*, D. Nejat, Editor. 2003, Academic Press. p. 3-14.

[147] Gregoriadis, G., R. Saffie, and J.B. De Souza, Liposome-mediated DNA vaccination. *FEBS Letters,* 1997. 402(2-3): p. 107-110.

[148] Mahato, R.I., Water insoluble and soluble lipids for gene delivery. *Advanced Drug Delivery Reviews,* 2005. 57(5): p. 699-712.

[149] Jiang, L., et al., Novel chitosan derivative nanoparticles enhance the immunogenicity of a DNA vaccine encoding hepatitis B virus core antigen in mice. *J. Gene Med.,* 2007. 9(4): p. 253-64.

[150] Duceppe, N. and M. Tabrizian, Advances in using chitosan-based nanoparticles for in vitro and in vivo drug and gene delivery. *Expert Opinion on Drug Delivery*, 2010. 7(10): p. 1191-1207.

[151] Nguyen, D.N., et al., Polymeric Materials for Gene Delivery and DNA Vaccination. *Advanced Materials*, 2009. 21(8): p. 847-867.
[152] Jilek, S., H.P. Merkle, and E. Walter, DNA-loaded biodegradable microparticles as vaccine delivery systems and their interaction with dendritic cells. *Advanced Drug Delivery Reviews*, 2005. 57(3): p. 377-390.
[153] O'Hagan, D.T., M. Singh, and J.B. Ulmer, Microparticles for the delivery of DNA vaccines. *Immunological Reviews*, 2004. 199(1): p. 191-200.
[154] Kim, H.K. and T.G. Park, Comparative study on sustained release of human growth hormone from semi-crystalline poly(-lactic acid) and amorphous poly(,-lactic-co-glycolic acid) microspheres: morphological effect on protein release. *Journal of Controlled Release,* 2004. 98(1): p. 115-125.
[155] Tinsley-Bown, A.M., et al., Formulation of poly(-lactic-co-glycolic acid) microparticles for rapid plasmid DNA delivery. *Journal of Controlled Release*, 2000. 66(2-3): p. 229-241.
[156] Anderson, J.M. and M.S. Shive, Biodegradation and biocompatibility of PLA and PLGA microspheres. *Advanced Drug Delivery Reviews,* 1997. 28(1): p. 5-24.
[157] Pack, D.W., *DNA delivery: Timing is everything.* Nat Mater, 2004. 3(3): p. 133-134.
[158] Elfinger, M., S. Uzgun, and C. Rudolph, Nanocarriers for Gene Delivery -Polymer Structure, Targeting Ligands and Controlled-Release Devices. *Current Nanoscience*, 2008. 4: p. 322-353.
[159] Lampela, P., et al., Effect of cell-surface glycosaminoglycans on cationic carrier combined with low-MW PEI-mediated gene transfection. *International Journal of Pharmaceutics*, 2004. 284(1-2): p. 43-52.
[160] Read, M.L., A. Logan, and L.W. Seymour, Barriers to gene delivery using synthetic vectors. *Adv. Genet.,* 2005. 53: p. 19-46.
[161] Gamvrellis, A., et al., Vaccines that facilitate antigen entry into dendritic cells. *Immunol. Cell Biol.,* 2004. 82(5): p. 506-16.
[162] Luten, J., et al., Biodegradable polymers as non-viral carriers for plasmid DNA delivery. *J. Control. Release*, 2008. 126(2): p. 97-110.
[163] Xiang, S.D., et al., Pathogen recognition and development of particulate vaccines: Does size matter? *Methods*, 2006. 40(1): p. 1-9.
[164] Alpar, H.O., et al., Biodegradable mucoadhesive particulates for nasal and pulmonary antigen and DNA delivery. *Adv. Drug Deliv. Rev.*, 2005. 57(3): p. 411-30.
[165] Xiang, S.D., et al., Delivery of DNA vaccines: an overview on the use of biodegradable polymeric and magnetic nanoparticles. Wiley Interdisciplinary Reviews: *Nanomedicine and Nanobiotechnology*, 2010. 2(3): p. 205-218.
[166] Guermonprez, P., et al., Antigen presentation and T cell stimulation by dendritic cells. *Annu. Rev. Immunol.*, 2002. 20: p. 621-67.
[167] Foged, C., et al., Particle size and surface charge affect particle uptake by human dendritic cells in an in vitro model. *Int. J. Pharm.,* 2005. 298(2): p. 315-22.
[168] Kanchan, V. and A.K. Panda, Interactions of antigen-loaded polylactide particles with macrophages and their correlation with the immune response. *Biomaterials*, 2007. 28(35): p. 5344-57.
[169] Dobrovolskaia, M.A. and S.E. McNeil, Immunological properties of engineered nanomaterials. *Nat. Nanotechnol.*, 2007. 2(8): p. 469-78.

[170] Fifis, T., et al., Size-dependent immunogenicity: therapeutic and protective properties of nano-vaccines against tumors. *J. Immunol.,* 2004. 173(5): p. 3148-54.

[171] Reddy, S.T., M.A. Swartz, and J.A. Hubbell, Targeting dendritic cells with biomaterials: developing the next generation of vaccines. *Trends Immunol.,* 2006. 27(12): p. 573-9.

[172] Manolova, V., et al., Nanoparticles target distinct dendritic cell populations according to their size. *Eur. J. Immunol.,* 2008. 38(5): p. 1404-13.

[173] Storni, T., et al., Nonmethylated CG motifs packaged into virus-like particles induce protective cytotoxic T cell responses in the absence of systemic side effects. *J. Immunol.,* 2004. 172(3): p. 1777-85.

[174] Mottram, P.L., et al., Type 1 and 2 Immunity Following Vaccination Is Influenced by Nanoparticle Size:,Äâ Formulation of a Model Vaccine for Respiratory Syncytial Virus. *Molecular Pharmaceutics*, 2006. 4(1): p. 73-84.

[175] Singh, M., A. Chakrapani, and D. O'Hagan, Nanoparticles and microparticles as vaccine-delivery systems. *Expert Rev. Vaccines,* 2007. 6(5): p. 797-808.

[176] Oster, C.G., et al., Cationic microparticles consisting of poly(lactide-co-glycolide) and polyethylenimine as carriers systems for parental DNA vaccination. *J. Control. Release,* 2005. 104(2): p. 359-77.

[177] Sloat, B.R., et al., Strong antibody responses induced by protein antigens conjugated onto the surface of lecithin-based nanoparticles. *Journal of Controlled Release,* 2010. 141(1): p. 93-100.

[178] Caputo, A., et al., Induction of humoral and enhanced cellular immune responses by novel core-shell nanosphere-and microsphere-based vaccine formulations following systemic and mucosal administration. *Vaccine,* 2009. 27(27): p. 3605-15.

In: DNA Vaccines: Types, Advantages and Limitations
Editors: E. C. Donnelly and A. M. Dixon
ISBN 978-1-61324-444-9


*Chapter III*

# Novel Delivery Strategies for DNA

***Jiafen Hu, Nancy Cladel and Neil Christensen***
Jake Gittlen Cancer Research Foundation, Pathology Department,
Pennsylvania State University College of Medicine, Hershey, PA 17033, USA

## Abstract

DNA vaccination has been dubbed the "third revolution" in vaccine development by some observers ever since its promising in vivo application in the early 1990s. The method is an attractive vaccine avenue because of its inherent simplicity as well as its relative cost effectiveness. DNA vaccination is also appealing because it is possible to produce large quantities of vaccine in a short period of time. In comparison with other conventional vaccines, DNA vaccines are stable and have a long shelf life. The most intriguing property of a DNA vaccine is that it can promote long-lasting humoral and cellular immunity leading to protective and/or therapeutic effects in preclinical and clinical tests. However, the immunogenicity of a DNA vaccine is hindered by suboptimal delivery of the DNA to cells, especially in large animals. Several other concerns have been raised in the field including the potential for DNA integration into the host genome and the possibility of autoimmune responses due to the long-term presence of the foreign DNA. In this review, we discuss different approaches to optimize DNA vaccines in order to improve their immunogenicity and delivery. Discussion will include the choice of the antigen, codon optimization to increase antigen expression, choice of optimal expression vectors and proper adjuvants; there will be a special focus on several novel delivery methods. We will also discuss the potential strategies to address the concerns in DNA vaccine application.

## I. Introduction

DNA vaccines consist of a DNA plasmid encoding the sequence of a target protein from a pathogen under the control of a eukaryotic promoter. Naked DNA vaccines are attractive non-viral vectors because of their inherent simplicity. They can easily be produced in bacteria and manipulated using standard recombinant DNA techniques. Some observers have already dubbed this new technology the "third revolution" in vaccine development—equivalent to Pasteur's ground-breaking work with whole organisms followed by the development of subunit vaccines in early 1990's [1]. Dr. Jon Wolff was the first to inject DNA plasmids into mice and to observe in situ protein production [2]. Later, Dr. Margaret Liu found that intramuscular injection of DNA from influenza virus into mice produced specific immune responses [3]. The feasibility of DNA immunization against a few key human pathogens, including influenza virus, human immunodeficiency virus (HIV), human hepatitis B virus (HBV) and herpes simplex virus (HSV) has been demonstrated in small animals [4-6]. DNA vaccination to induce immunity against certain cancers has also been demonstrated [7-10]. The first clinical trials using injections of DNA to stimulate an immune response against a foreign protein began for HIV in 1995. Four other clinical trials using DNA vaccines against influenza, HSV, T-cell lymphoma, and an additional trial for HIV followed [6, 11-13]. According to the *Journal of Gene Medicine* there are currently 246 clinical trials worldwide investigating the utility of naked plasmid DNA vaccination (www.wiley.co.uk/genmed /clinical). This database represents almost 20% of all vectors being used in clinical trials and comes in third place behind adenovirus (24.8%) and retrovirus (22.3%) vectors.

DNA vaccines have several advantages over conventional vaccines such as attenuated live vaccines, inactive whole viral vaccines, subunit vaccines, recombinant protein vaccines and peptide vaccines [8, 14-16]. First, the most attractive property of DNA vaccination is its ability to stimulate strong humoral and cell-mediated immune responses that can be both prophylactic and therapeutic [9, 17]; Second, DNA vaccines are designed to target the antigen of interest and the immune responses are specific for targeted cells in the hosts, and thus very few side effects are expected [15]; Third, contrary to common belief, long-term foreign gene expression from DNA vaccines have been demonstrated even in the absence of integration into the chromosome when the target cell is postmitotic (as in muscle) or slowly mitotic (as in hepatocytes) [18-20]. Therefore, long-term protective/ therapeutic outcomes following DNA vaccination becomes possible; Fourth, DNA vaccines can be conveniently and cheaply produced and purified and therefore can be made available to developing countries [7, 21]; Fifth, DNA vaccines have a long shelf life in comparison with other conventional vaccines and refrigeration is not required for transportation or storage [14]; Sixth, the amount of time necessary for the development of DNA vaccines is relatively short and this should enable the timely production of vaccines against emerging infectious diseases; Seventh, DNA vaccines show very little dissemination and transfection at distant sites following delivery and can be re-administered multiple times to mammals (including primates) without inducing an antibody response against itself (i.e., no anti-DNA antibodies are generated).

This revolutionary technology has proven to be clinically effective and four DNA vaccines have recently been approved for veterinary use [22, 23]. DNA vaccines against bacterial infections have shown encouraging results [24]. DNA vaccines have been tested for the prevention and treatment of allergies as well as cancers [25-27]. For example, prophylactic and therapeutic DNA vaccines against papillomavirus-induced cervical cancer have been studied extensively in the last two decades [26, 28, 29]. Because of the species specificity of HPV, no animal models are available to study HPV directly. However, several naturally occurring animal papillomavirus models have played a pivotal role in the development of effective vaccines for HPV infection. These vaccines include Gardasil and Cervarix, now commercially available. In addition, these preclinical animal models have been used to test the immunogenicity of different viral antigens [30-33]. Prophylactic and therapeutic DNA vaccines have been tested in two naturally occurring infectious animal papillomavirus models: canine papillomavirus [34, 35] and rabbit papillomavirus models [36-46]. In addition to the whole antigen, an array of epitopes targeting B and T cell mediated immune responses has been identified and tested [47, 48]. Epitope DNA vaccines from these epitopes have shown promising results [49-54].

Despite the advantages and potential of DNA vaccination, immune responses by DNA vaccination were suboptimal in early studies. The weak responses were due in part to the limitations of delivery, uptake and protein expression from the DNA vaccine within the cells. Since those early studies, every aspect of DNA vaccines including the selection of the expression plasmid, codon usage of the antigen, delivery techniques and adjuvants has been investigated extensively to improve vaccine immunogenicity (Figure 1) [13, 55]. In this review, we will discuss some of these optimization strategies in detail.

# II. Optimization of Immunogenicity of DNA Vaccines

## 2.1. Choice of Targets and Expression Plasmid for DNA Vaccines

For a given DNA vaccine, the first and most important consideration is to choose effective targets.

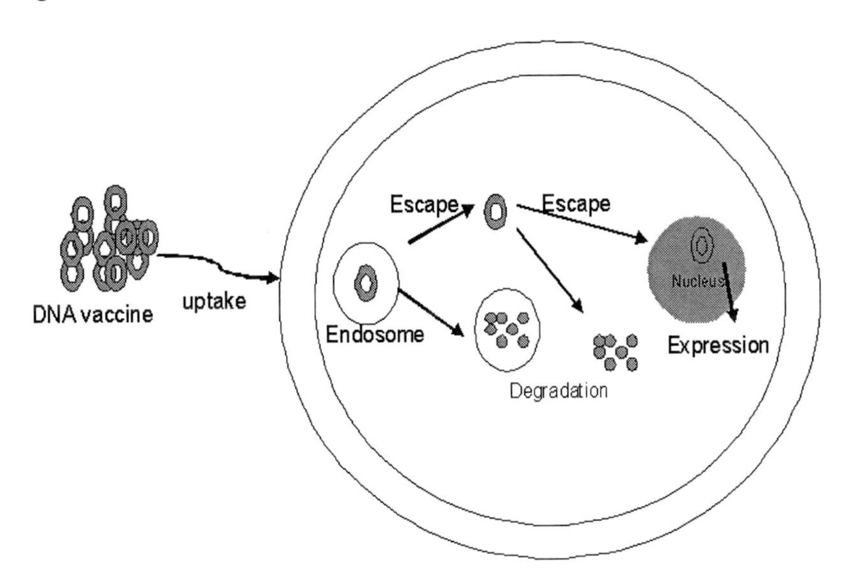


Figure 1. Schematic diagram of DNA delivery pathways showing three major barriers: low uptake across the plasma membrane, inadequate release of DNA molecules with limited stability, and lack of nuclear targeting. DNA is taken up into the cells by endocytosis. Some of the DNA escapes from the endosome while most is degraded inside the endosome. A few copies of the escaped DNA are then released intracellularly and later enter into the nucleus for expression. Most of the escaped DNA, however, is degraded in the cytosol. Adapted from Luo D. and Saltzman W. M with some modifications, NATURE BIOTECHNOLOGY VOL 18, 2000.

DNA vaccination has been found to induce both humoral and cellular immune responses in vivo and subsequently to be protective and/or therapeutic. It is critical to identify the "right" targets to optimize protective host immunity to successfully eliminate a given pathogen [16].

Influenzavirus surface antigens such as hemagglutinin (HA) or neuraminidase (NA) have been used as vaccines to stimulate protective immunity against influenza virus [12, 55, 56]. However, the frequent mutations of the HA gene necessitates redesign of the vaccine every year and this has posed significant challenges. Recent studies have demonstrated that nucleoprotein (NP) as well as the M gene, segment 7 of the influenza virus gene, when formulated as DNA vaccines can enhance cell-mediated immunity. Because limited mutations occur in these genes, they are promising targets for next generation DNA vaccines for influenzavirus infections [57,58].

Herpes simplex virus (HSV) gD2 as a practical vaccine antigen has been evident for almost 30 years because anti-gD antibodies have potent neutralizing activity [59, 60]. More recently, CD4-and CD8-specific epitopes in gD2 have also been defined [6, 59, 61, 62]. The amino acid sequences of gD1 and gD2 are highly homologous, as are the principle cellular receptors used by gD of HSV-1 and HSV-2 to facilitate viral entry. An effective gD-based DNA vaccine thus might provide cross-protection against both HSV-1 and HSV-2 [6, 59].

The human papilloma virus (HPV) genome contains seven early genes and two late genes. These genes have been found to be expressed in different cells at different levels as well as at different stages of disease (Figure 2A) [63]. Two early oncoproteins, the E6 and E7, form ideal targets for therapeutic HPV vaccines, since they are constituitively expressed in HPV-associated cervical cancer and its precursor lesions and thus play crucial roles in the generation and maintenance of HPV-associated disease (Figure2B ) [63]. We and others have made use of this knowledge to design preventive and therapeutic DNA vaccines from different early and late genes using natural infection preclinical models [34-39, 41-43, 64]. Interestingly, L1, which has been developed clinically as a prophylactic vaccine, is capable of inducing cell mediated immune responses when applied as a DNA vaccine although no detectable L1 expression is found in basal layers of infected tissues [39,65-67].

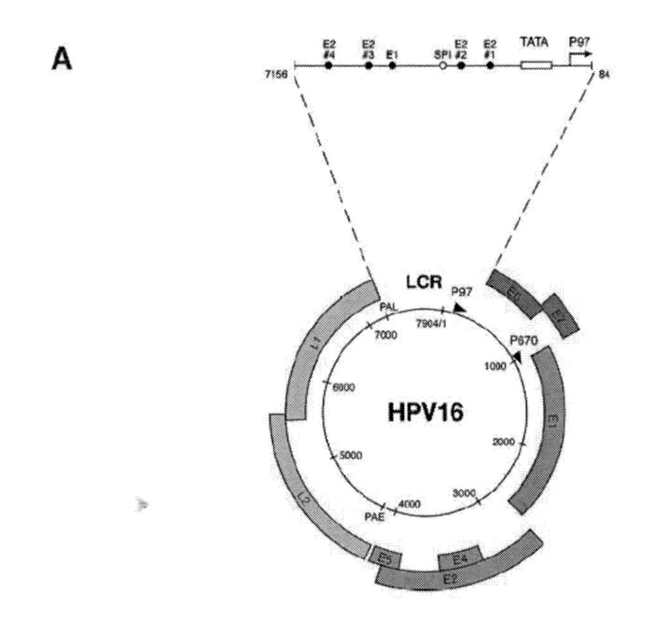

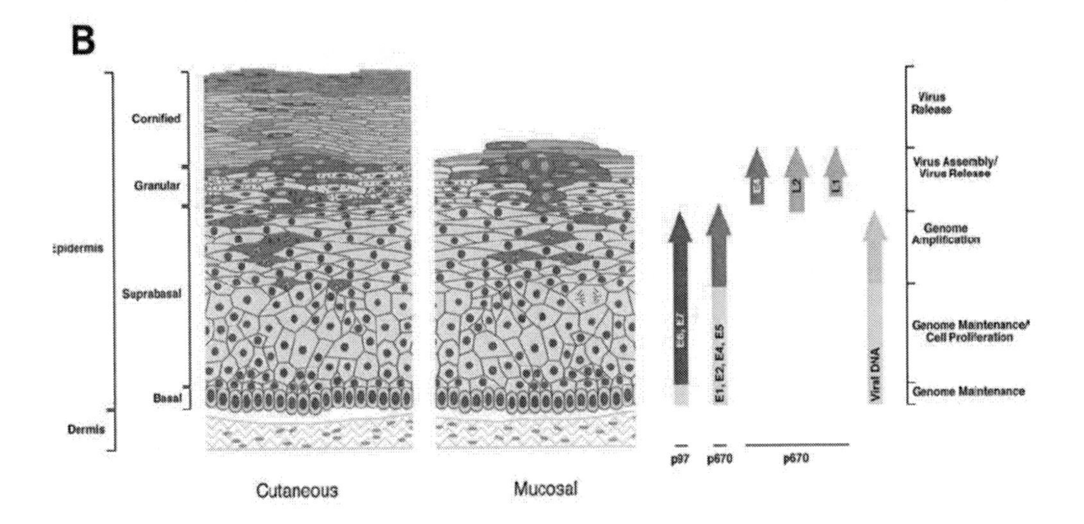


Figure 2. Organization of the HPV genome (A) and the virus life cycle (B). From John Doorbar, Molecular biology of human papillomavirus infection and cervical cancer. Clinical Science (2006) 110, 525–541.

Earlier studies demonstrated that vaccines targeting multiple antigens or epitopes provided the strongest protective and therapeutic immunity [46, 54, 55, 68-76]. Two recently completed phase I clinical trials with candidate HIV-1 vaccines demonstrated that DNA vaccines are, indeed, immunogenic in humans, even when administered through routine needle injections [10].

One major challenge to immune-induced clearance of diseases is that immunodominant epitope-induced T cell exhaustion and tolerance can occur, especially in cancer immunotherapy [27]. Understanding the immunological mechanisms that occur during infections and cancer would greatly help in the screening of better targets for DNA vaccine design [77]. Recently, we established a novel human MHCI (HLA-A2.1) transgenic rabbit to facilitate the development of therapeutic vaccines against HPV induced cancers [75, 78] and ocular HSV [62]. Taking advantage of online prediction software to search for specific epitopes for this MHCI, we have been able to screen and test specific HPV targets for improved HPV DNA vaccine design [78]. A multivalent epitope DNA vaccine strategy was effective for both protective and therapeutic immunity in our studies as well as in other reports [27, 54, 55, 75]. Our most recent work demonstrated that the immune responses from a multivalent vaccine with these epitopes were not uniform in all animals tested (Figure 3). Whether this variable immunity from a multivalent DNA vaccine could achieve a better outcome than a combination of single epitope DNA vaccines remains to be determined [70].

## 2.2. Selection of High Expression Plasmids for DNA Vaccines

A large array of highly efficient expression vectors has been developed to increase expression from DNA vaccines (Figure 4) [13, 16, 23, 79, 80]. The most commonly used expression vectors contain either the cytomegalovirus (CMV) or Rous sarcoma virus (RSV) promoters [14].

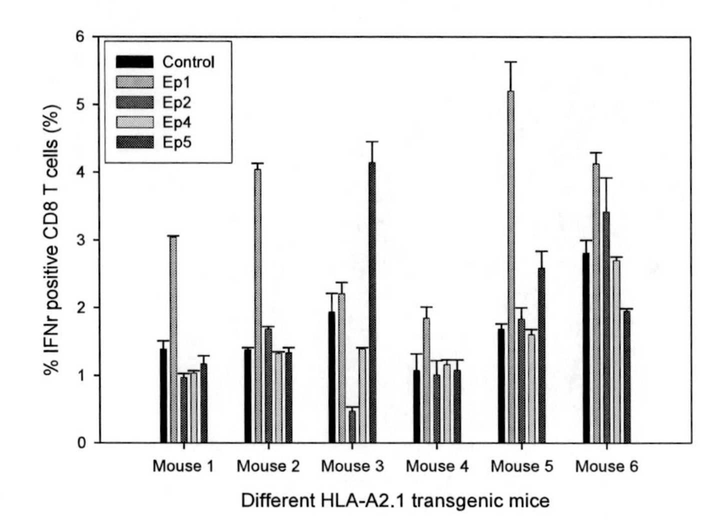


Figure 3. CRPVE1multivalent epitope DNA vaccination in HLA-A2.1 transgenic mice. Six 6-8 week old HLA-A2.1 transgenic mice were immunized with a CRPV E1 multiepitope DNA vaccine by gene-gun for two times with a two-week interval between vaccinations. The animals were sacrificed and spleens were harvested for in vitro stimulations. After two in vitro stimulations with the corresponding peptide, the bulk CTLs were analyzed for intracellular interferon gamma (IFNγ) gated on CD8 T cells. Ep1 located at the first position in the vaccine was recognized by five of the six animals. Therefore, it is possible that the position of the epitope plays a role in determining the immunogenicity of the epitope.

These promoters, however, have been found to be attenuated by DNA vaccine-induced cytokine signals in vivo. Recently, a constitutive cellular promoter (beta-actin) has been found to less sensitive to cytokine mediated repression of the transgene expression [81].

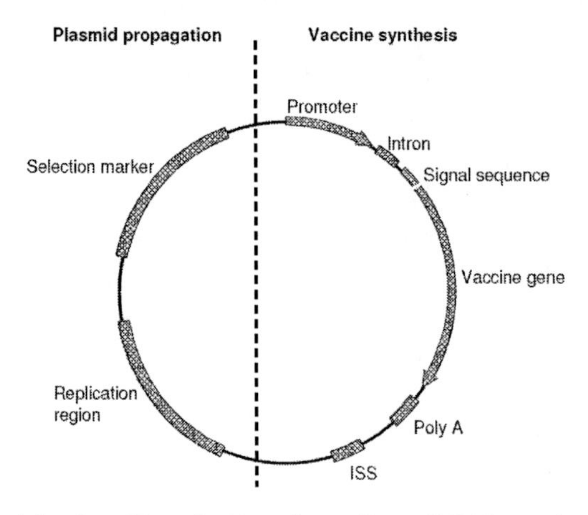


From Jacob Glenting and Stephen Wessels, Ensuring safety of DNA vaccines. Microbial Cell Factories 2005, 4:26.

Figure 4. Genetic elements of a plasmid DNA vaccine. Plasmid DNA vaccines consist of a unit for propagation in the microbial host and a unit that drives vaccine synthesi s in the eukaryotic cells. For plasmid DNA production a replication region and a selection marker are employed. The eukaryotic expression unit comprises an enhancer/promoter region, intron, signal sequence, vaccine gene and a transcriptional terminator (poly A). Immune stimulatory sequences (ISS) add adjuvanticity and may be localized in both units.

Promoters of the DNA plasmid are often methylated and this precludes optimal gene expression. Demethylating agents such as, 5-aza-2'-deoxycytidine can be combined with DNA vaccines to overcome this problem [82].

## 2.3. Codon Optimization for Higher Gene Expression

A number of pathogens including some DNA viruses use suboptimal codons (rare codons relative to those of the hosts) in order to escape immune surveillance [53, 83-86]. To improve the immunogenicity of DNA vaccines, a strategy of codon optimization has been utilized to increase protein expression levels of these target products in vitro and in vivo [23, 87-91]. The increased target antigens resulting from vaccination with codon-optimized DNA led to increased immunity and subsequent higher protection rates in hosts [23, 88, 92-94]. This strategy can be used to develop highly efficient prophylactic and therapeutic DNA vaccines. Specifically inactivating negative regulatory RNA elements is also an effective strategy to enhance gene expression [89]. While codon optimization has proven very useful in many cases, it should be noted that the rules governing codon usage are still imperfectly understood. In light of this fact, codon "optimizations" do not always have the desired effect and this possibility should not be ignored [91, 95, 96].

## 2.4. Adjuvant Choice for DNA Vaccines

Adjuvants have been used to augment DNA vaccine-induced immunity in numerous studies. For example, cytokines have been widely applied to augment the immunogenicity of DNA vaccines and this topic has been well reviewed in several other publications [4, 97, 98]. Microbial infection induces innate immunity by triggering pattern-recognition receptors including toll-like receptors (TLRs). The infected cells produce proinflammatory cytokines that directly combat microbial invaders and express costimulatory surface molecules; this then leads to the development of adaptive immunity by inducing antigen-specific T cell differentiation. For example, DNA vaccines administered together with a cytokine adjuvant such as IL-12 or GM-CSF enhances both Th1 and Th2 responses, respectively in preclinical models as well as in clinical trials [57, 99, 100].

CpG–DNA is another promising adjuvant when administered together with DNA vaccines [98, 102-104]. Unmethylated CpG motifs are relatively common in bacterial DNA, but are rare in mammalian and plant DNA. It is possible that they represent an evolutionary adaptation to augment innate immunity, most likely in response to pathogens that replicate within the host cells, such as viruses and intracellular bacteria [101]. Therefore, CpG-DNA functions as an adjuvant for regulating the initiation of Th1 differentiation and for the production of Thl cytokines as well as TNF-alpha [98, 102-104].

# III. DNA Delivery Strategies

Repeated studies have demonstrated that the immunogenicity of DNA vaccines greatly depends upon the delivery method used for immunization [10, 98, 105-108]. In the past decade, numerous papers have been published describing the barriers to effective DNA delivery (Figure 1). These barriers include 1) physical and biochemical degradation in the extracellular space. For example, DNA vaccines are high molecular weight molecules that do not passively enter intact skin; 2) internalization after entering into the cells; 3) trafficking from endosomes to lysosomes and subsequent degradation; 4) escape from endosomes into the cytoplasm; 5) dissociation of DNA from its carrier; 6) transfer into the nucleus, and 7) transcription (for gene transfer) or hybridization (for antisense inhibition). For in vivo applications, physicochemical properties of vectors that affect stability as well as interactions with extracellular systemic components and the immune system are also of significant importance. It would be desirable to identify a single step as the rate-limiting barrier. However, any or all of these processes may pose a significant hindrance to the effectiveness in a given application [109].

Because plasmid DNA is rapidly eliminated from the bloodstream owing to extensive uptake by the liver and kidney and because it exhibits poor cellular uptake owing to the large size and the highly anionic nature of DNA, low levels of gene expression are generally achieved [16]. Therefore, enthusiasm for DNA vaccination in humans is still tempered by the fact that optimal delivery of the DNA to cells has yet to be achieved [105]. Several delivery methods (systems) have been used for DNA vaccination. These include techniques such as needle injection, electroporation, topical application, and particle-mediated delivery via gene gun. They also include vehicle-associated delivery systems, both viral and non-viral. Some of these methods have shown promise in promoting strong protective and therapeutic immunity in preclinical and clinical models.

## 3.1. Invasive Delivery

### *3.1.1. Needle Injection*

DNA vaccine was initially administrated by conventional needle injection in early studies. Although subcutaneous or epidermal sites were occasionally targeted, muscle tissue was the primary target site for needle injection (Figure 5). Upon injection into the host's muscle tissue, the DNA is taken up by host cells, which then begin to express the foreign protein [16]. Some studies have demonstrated antibody generation and measurable protection against various infections by DNA vaccination [110]. Our earlier studies showed that while intramuscular injection of DNA vaccines did induce T cell mediated immune responses in vaccinated animals, these animals failed to gain any protection demonstrating that the immunity was suboptimal [36]. Over time, needle injection has been found to be the least effective method and usually requires high doses of DNA plasmid to achieve a measureable response [105]. DNA vaccines delivered by this technique has been tested in humans and found to be poorly immunogenic in some studies [10, 11, 13, 111, 112]. The possibility of cross contamination by repeated needle injection would be a potential problem.

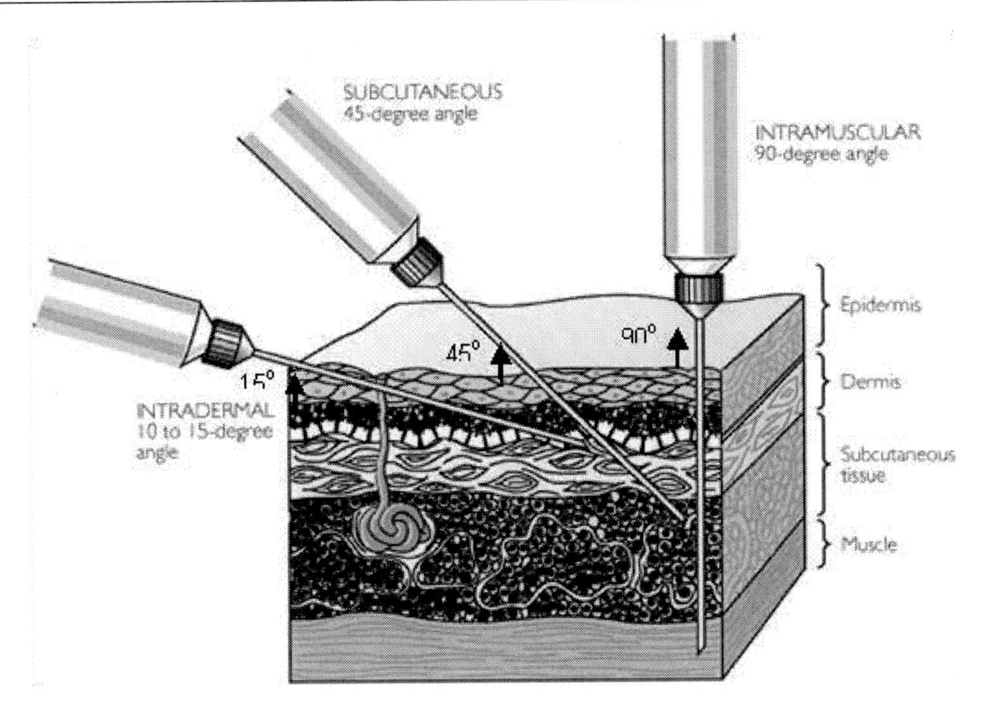


Figure 5. Needle injection routes for DNA vaccines. Plasmid DNA vaccines can be injected intramuscularly (with a 90° angle), subcutaneously (with a 45° angle) or intradermally (with a 15° angle) to target different cells for expression.

### *3.1.2. In Vivo Electroporation*

The technique of eletroporation has been used for over 25 years as a means of introducing macromolecules, including DNA, into cells *in vitro*, and is now widely used for transfection of plasmids into different tissues *in vivo* [113].

Electroporation has shown great promise as a delivery method for DNA vaccine administration [20, 107, 114-116]. This delivery method has proven more efficient than intramuscular injection in eliciting antigen-specific immunity even in large animals which is potentially useful for clinical application [116, 117]. However, significant tissue damage related to harsh electroporation conditions raises serious safety concerns with the use of electroporation in healthy tissues and thus limits its current applications to non-healthy tissues such as tumors. DNA formulations designed to minimize tissue damage or enhance expression following weaker electric pulses have been examined in an attempt to address these concerns. These strategies include formulations fortified with the addition of transfection reagent(s), membrane-permeating agents, tissue matrix modifiers, targeted ligands, or agents modifying electrical conductivity or membrane stability to enhance delivery efficiency or reduce tissue damage. Advancements in DNA formulation could improve the safety of electroporation protocols for human applications [117, 118].

### *3.1.3. Tattoo Gun Delivery*

The tattoo gun is one kind of electroporation that has shown promise in recent studies [119-121]. Instead of intramuscular injection, the DNA is applied to the surface of the skin and a five-needle unit oscillating at a voltage of 17.4 V and corresponding to a frequency of

145 Hz (145 punctures per second) is utilized. The tattoo device is adjusted to allow exposure of the needle tip 1–2 mm beyond the barrel guide. This depth of tattooing into mouse skin has been shown to result in the immediate location of tattooed inks mainly in the dermis and to a lesser extent in the epidermis [122]. The tattoo procedure causes many minor mechanical injuries followed by hemorrhage, necrosis, inflammation, and regeneration of the skin and thus non-specifically stimulates the immune system [123, 124]. Therefore, tattooing may partially substitute for the function of adjuvants as well as acting as a delivery vehicle for the DNA. As tattooing involves a much larger area of the skin than intradermal or intramuscular injection, it offers the advantage of potentially transfecting more cells [122]. Gene expression after DNA tattooing has been shown to be comparable to or higher than that following both intradermal injection and gene gun delivery (our unpublished observation).

Advantages of the tattoo device are the low price and the standardized method for the application [123]. DNA vaccines delivered by tattoo gun were able to induce both cellular and humoral antigen-specific responses (our unpublished observations) [122]. The primary disadvantages of tattoo delivery are the strain on the animals and the somewhat cumbersome application procedure. In particular, the local trauma induced by the tattooing procedure might be considered unacceptable in routine prophylactic vaccination settings involving human subjects [124]. Nevertheless, DNA vaccination via tattoo gun appears to be the method of choice if more rapid and stronger immune responses need to be achieved. Potential applications might be vaccination of livestock for prophylaxis or of human beings for therapeutic purposes.

## 3.2. Non-Invasive Topical Delivery Method

### *3.2.1. Tape Stripping*

The skin is the largest and the most accessible body organ and plays a key role in protection against pathogens. It acts as an efficient physical and immunological barrier. The specific immunologic environment of the skin, known as the skin associated lymphoid tissue (SALT), consists mainly of 1) Langerhans cells and dermal antigen-presenting cells which circulate between the skin and the lymph nodes, 2) keratinocytes and endothelial cells which produce a wide range of immune and growth regulatory cytokines and 3) lymphocytes which extravasate from the circulation into the skin [125]. The SALT provides both innate and adaptive immunity to protect the individual against microbial attack [126]. These immunological properties make the skin an attractive organ for the delivery of vaccines [76, 127]. However, due to the skin's barrier properties, the penetration of DNA and the applications of topical vaccination are limited.

To improve permeability of the stratum corneum and the potency of topical DNA vaccines, efficient delivery systems are needed. Topical vaccination has been achieved using topical application of naked DNA in conjunction with tape stripping. Tape stripping has been found to activate langerhans cells and keratinocytes [128]. Tape stripping also induces expression of TLR 9 and is an effective way to induce a Th1-type immune response after topical application of CpG-ODN and antigen [108]. Tape stripping was also shown to exert an interesting adjuvant effect for DNA vaccines following electroporation [129].

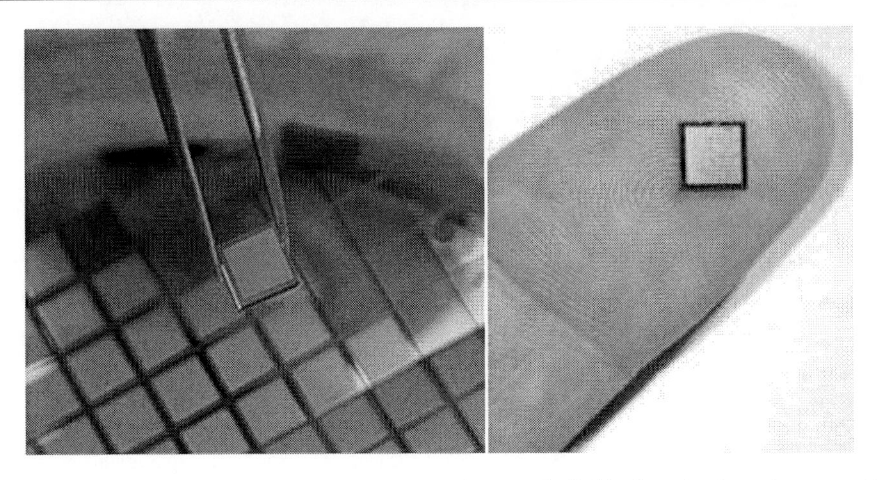

Figure 6. Nanopatch (NP) vaccine delivery. NPs contain 3364 individual projections that were 30mm at the base and between 65 and 110mm in length. These micro projections were coated with a novel nitrogen-jet drying method that resulted in a consistent and robust layer of DNA. The coated NP was applied to the skin at 2.0+0.1ms-1 for 10 min. After NP removal, the coating that was on the micro projections appears to have remained in the skin as expected. During NP application the skin is penetrated by the projections and the strata compressed at the puncture site. The penetration of the skin by the coated micro projections resulted in the delivery of antigens to the epidermis and the upper-dermis.

Previous studies on so-called epicutaneous vaccination demonstrated that tape stripping facilitated percutaneous penetration of antigens applied onto the skin and modulated antigen-specific immune responses when co-administered with Cholera toxin or CpG ODN [130]. A recent publication showed that stratum corneum disruption was a feasible and well tolerated procedure in humans and therefore a promising approach for clinical application of DNA vaccines [10, 111].

### *3.2.2. Nanopatch Delivery*

Several studies have reported a new non-invasive micro projection array delivery method [131] (Figure 6). This method has been used successfully to deliver DNA vaccines in vivo and to induce strong protective immunity in vaccinated animals [131-133]. The patch uses 100 times less vaccine than what is needed to be delivered via direct injection to provide the same immunity.

### *3.2.3. Nano-Particle Mediated Delivery*

The gene gun is a biolistic (contraction of biological and ballistic) device that enables DNA delivered under high pressure to directly transfect keratinocytes and epidermal langerhans cells (immature dendritic cells, iDC) with minimal damage to the skin [134, 135]. These events stimulate DC maturation and migration to the local lymphoid tissue, where DCs prime T cells for antigen–specific immune responses [136]. Traditionally, gene gun technology has been particle-mediated and involves using compressed helium to propel a stream of gold particles coated with DNA into the skin, where langerhans cells are located [137]. The gene gun is currently widely used in many immunotherapeutic applications in preclinical studies [138]. The delivery of DNA vaccines by gene gun was shown to be the most dose-efficient method of vaccine administration in comparison with routine intramuscular and biojector injection and achieved similar immune responses as those elicited by the electroporation method [139-141].

DNA vaccines delivered by gene gun have proven highly protective and therapeutic in the mouse model as well as in several preclinical papillomavirus infection models [34, 35, 37-39, 41, 44, 45, 75, 99, 142-145]. Gene-gun delivery is relatively non-invasive. However, it is very costly (requiring pure gold particles for delivery of the DNA vaccine) and is not very practical (requires specialized equipment) in the clinical setting [10]. Recent advances in gene gun technology have paved the way for non-particle mediated delivery of DNA in solution under low pressure. Both protective and therapeutic effects have been observed and there is fewer traumas to the skin relative to high pressure particle mediated delivery [146]. Further studies will reveal whether this less invasive tool will have broad applicability.

### 3.3. Viral-Mediated DNA Delivery Systems

Viral-mediated DNA delivery has shown recent promise [147]. Foreign DNA was encapsidated within the capsids of the influenza virus as virosomes. These virosomes retained the membrane fusion properties of the native influenza virus and were highly efficient vehicles for DNA delivery. The virosomes provided complete DNA protection from nuclease degradation, which is especially important for future in vivo applications [148].

DNA can be encapsidated inside other viral capsids such as the papillomavirus L1/L2 capsid as a "pseudovirus" to mimic natural infection [105, 124, 147, 149, 150]. Papillomavirus capsid protein L1 or in combination with L2 can assemble into virus-like particles (VLPs) in vitro and these particles have the capacity to encapsidate unrelated plasmid DNA [151-153]. In addition to delivering encapsidated DNA, HPV VLPs have been shown to effectively deliver externally attached DNA in vitro and in vivo [154]. Our studies have confirmed the external delivery of DNA by capsids, have shown that the delivery is L2 dependent and have demonstrated that the efficiencies differ from virus to virus (Figure 7). The efficiency of externally attached DNA delivery is type specific. For example, capsids of HPV58 showed significantly higher efficiency than those of HPV16, the capsids that are used for most reported studies. In addition, externally attached DNA delivery has been shown to be more efficient than encapsidated DNA delivery indicating the possibility that more copies of DNA can be attached to VLPs than are encapsidated within them (unpublished observations).

Recombinant adenoviral vectors have served as one of the most efficient gene delivery vehicles in vivo thus far [22, 155]. Multiply attenuated adenoviral vectors have been developed to achieve long-term gene expression in animal models by overcoming cellular immunity against de novo synthesized adenoviral proteins [155]. Adenoviruses overcome a series of biological barriers, including endosomal escape, intracellular trafficking, capsid dissociation, and nuclear import of DNA, to deliver their genome to the host cell nucleus. Adenoviruses have been used for DNA delivery in several studies [1, 7, 11, 21]. The Adenovirus delivery system has several advantages for DNA delivery in future clinical use because it has 1) a large host range; 2) a low pathogenicity in humans; 3) a capacity for including inserted foreign DNA (>30 kb); and 4) the possibility to obtain high titers of virus, which is important for in vivo applications.

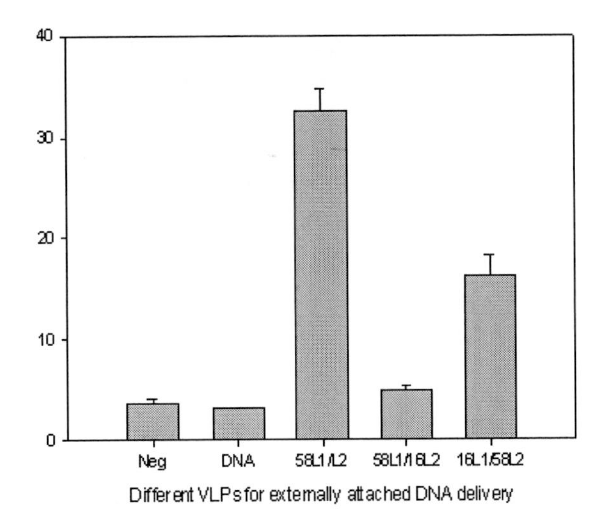


Figure 7. Delivery efficiency of externally attached DNA is L2 dependent. Hybrid VLPs (HPV16L1/58L2 and HPV58L1/HPV16L2) were tested for DNA delivery into RCLT Although neither was as efficient as HPV58L1/L2 VLPs, significantly more GFP positive cells were found by HPV16L1/HPV58L2 mediated DNA delivery than by HPV58L1/HPV16L2 delivery.

However, repeated boosters by viral-mediated DNA delivery systems are not optimal because of the interference of humoral immunity against the viral capsid. In addition, the undefined risk of infection or inflammation could compromise the use of this DNA delivery option [105].

## 3.4. Non-Viral Gene Delivery Systems

Non-viral gene delivery methods are gaining recognition as an alternative to viral gene vectors due to their potential to avoid immunogenicity and toxicity problems inherent in the use of viral systems [156-158]. Naked plasmid DNA is a powerful tool for gene therapy, but it is rapidly eliminated from the circulation after intravenous administration. Solid lipid nanoparticles (SLNs) have demonstrated transfection capacity in vitro and effective delivery of DNA for expression in vivo [159].

Polymeric microparticles 1–2 um in diameter and consisting of Poly (D, L) Glycolic-Co-Lactic Acid (PGLA), a biocompatible polymer, are used in a number of pharmaceutical products, including sutures. These microparticles have a greater propensity to be taken up by antigen-presenting cells (such as macrophages and dendritic cells) relative to naked DNA. This technique allows DNA plasmids to be condensed inside the microparticles. The physical and chemical properties of the PGLA scaffold make DNA inaccessible to nucleases. This latter event will prevent DNA degradation and allow for a sustained release of DNA [156, 160].

Many cationic polymers have been studied both in vitro and in vivo for gene delivery purposes [161]. However, in recent years there has been a focus on biodegradable carrier systems [158]. The potential advantage of biodegradable carriers as compared to their non-degradable counterparts is the reduced toxicity and the prevention of accumulation of the polymer in the cells after repeated administration. Also, the degradation of the polymer can be used as a tool to release the plasmid DNA into the cytosol [158, 162, 163].

Liposomes have emerged as one of the most versatile tools for the delivery of DNA therapeutics. Liposomes are vesicles that consist of an aqueous compartment enclosed in a phospholipid bilayer. Multiple types of liposomes including cationic, anionic, and PH-sensitive have been investigated for DNA delivery over recent years [111, 164]. Several reviews have discussed this delivery system [156, 162, 165].

## 3.5. Novel Delivery Methods

### *3.5.1. Laser Mediated Delivery*

The femtosecond infrared titanium sapphire laser beam has been developed specifically for enhancing in vivo gene delivery without risks of tissue damage. A laser beam can deliver a focused amount of energy onto a target cell, thereby modifying permeability of the cell membrane at the site of the impact by a local thermal effect in vitro. This transient perturbation is sufficient to allow DNA in the surrounding medium to be transferred into the cells [5]. More recently, Tsen et al. found this novel technology to be an effective method for enhancing the transfection efficiency of intradermally injected plasmids [166]. Intradermal administration of DNA vaccines followed by pulses of laser has been performed successfully and has induced antigen-specific CD4 and CD8 T cell immune responses as well as humoral immunity [167]. Intradermal administration of HPV DNA vaccines followed by laser treatment has been demonstrated to be effective [168]. Based upon a limited number of studies, this novel technology appears to hold a high potential for delivery of a therapeutic DNA vaccine.

### *3.5.2. Ultrasound Mediated Delivery System*

Ultrasound mediated delivery is noninvasive, has the ability to focus energy deep within the body, has the ability to precisely target the tissue region of interest while leaving intervening tissues unaffected, and is non-ionizing, allowing repeated applications without limitations imposed by dose. This delivery system has shown promise in increasing uptake of drugs via a process known as sonoporation [169]. Sonoporation increases the permeability of cell walls and therefore allows the effective uptake of large molecules such as DNA [170-172]. The ultrasound-mediated technique shows no risk of side effects such as anti-viral immune or inflammatory responses [173-175]. In addition, it is easier to apply and more cost-effective than some systems, especially when compared with the gold particle mediated DNA delivery that we have used in our immunization studies.

Ultrasound has been shown to transiently disrupt cell membranes and, thereby, facilitate the loading of drugs and DNA into viable cells [173, 176, 177]. Previous studies demonstrated that optimization of physical parameters could significantly increase the delivery of DNA plasmid to cells in vitro and in vivo [175, 178-180]. The Ultrasound-mediated micro-bubble delivery has been tested for delivery of DNA and siRNA in preclinical animal models and has been found to be effective but inefficient [181, 182]. The critical roadblock to date has been that very low numbers of cells that took up DNA plasmid into the cytosol actually showed expression [176, 183]. These results suggest that further optimization of the ultrasound technique is needed.

# III. Future Directions

Despite the promising potential of DNA vaccination, the immune responses from DNA vaccines need to be improved utilizing the strategies discussed in this review as well as others yet to be developed. The delivery, uptake and presentation of DNA to cells is still not optimal, particularly in larger animals [10, 16, 20, 80, 98, 105, 184]. New strategies such as the combination of DNA vaccines with other vaccines including peptide, protein and recombinant viruses as well as novel adjuvants and different delivery techniques have been studied [107, 185-188]. These new strategies will be expected to augment the immunogenicity of DNA vaccines against pathogens as well as tumors in future clinical applications.

Immunization times and intervals have been demonstrated to be important parameters that can significantly influence vaccine effectiveness. Earlier studies have demonstrated a requirement for booster immunizations to achieve the highest level of protection and therapeutic effect [189]. In our model system, we showed that one DNA booster immunization was required and sufficient to promote protective immunity comparable to that provided by two booster DNA immunizations [75]. Other published studies showed that increasing the interval between immunizations significantly enhanced the frequency and magnitude of CD8+ and CD4+ T cell responses as well as protective immunity [190]. In contrast, another study demonstrated that repeated DNA immunizations with short intervals between boosters promoted strong therapeutic effects of the DNA vaccine in a mouse tumor model [191]. Therefore, different model systems may have their own optimal parameters to achieve the highest protection.

Several safety concerns have been raised for DNA vaccinations. The first concern is one which exists with all gene therapy, which is that the DNA vaccine may be integrated into host chromosomes and potentially activate oncogenes or deactivate tumor suppressor genes [19, 192]. Although several studies have demonstrated negligible DNA vaccination-induced mutations and integrations which may help to dispel this concern, it will take years to understand and possibly resolve the potential problem [192-196]. The second concern is that extended immunostimulation by prolonged expression of the foreign antigen could, in theory, provoke chronic inflammation or autoantibody production [12, 194, 195, 197]. This possible side effect can be potentially reduced by using epitope DNA vaccines. However, more basic research is needed to understand the long-term effects of DNA vaccination. The third concern is that selectable antibiotic markers have been used to engineer plasmid DNA vaccines for most studies [198], and these genes are also delivered into host tissues. Although this is a powerful selection method for in vitro manipulation of plasmids, genes which confer resistance to antibiotics are discouraged by regulatory authorities [80, 194]. The main concern is that the plasmid may transform the patient's microflora and spread the resistance genes. A non-antibiotic-based marker on vaccine plasmids for use in *E. coli* has been developed. This system is based on the displacement of repressor molecules from the chromosome to the plasmid, allowing expression of an essential gene [199]. Such a selection system is efficient and precludes the use of antibiotics. Finally, the immunogenicity of a DNA vaccine is dependent upon proper processing of the antigen and this may not always occur. Therefore DNA vaccination is not necessarily a universal strategy for every antigen.

In summary, DNA vaccination is a promising strategy for both protective and therapeutic applications. In contrast to conventional vaccines, DNA vaccines are cost-effective, easy to administer and can be made in large quantities within a short period of time. In both preclinical and clinical experiments, DNA vaccines have been shown to evoke both humoral and cellular immune responses. With the optimization of all aspects from vaccine design to delivery systems (methods), DNA vaccinations will continue to be a valuable approach to combat infectious diseases and cancers.

# References

[1] Gao L, Wagner E, Cotten M, et al. Direct in vivo gene transfer to airway epithelium employing adenovirus-polylysine-DNA complexes. *Hum.Gene Ther.* 1993;4:17-24.
[2] Rooney JF, Bryson Y, Mannix ML, et al. Prevention of ultraviolet-light-induced herpes labialis by sunscreen. *Lancet* 1991;338(8780):1419-22.
[3] Donnelly JJ, Ulmer JB, Liu MA. Protective efficacy of intramuscular immunization with naked DNA. *Ann.N.Y. Acad.Sci.* 1995;772:40-6.
[4] Lori F, Weiner DB, Calarota SA, Kelly LM, Lisziewicz J. Cytokine-adjuvanted HIV-DNA vaccination strategies. *Springer Semin. Immunopathol.* 2006;28(3):231-8.
[5] Zeira E, Manevitch A, Manevitch Z, et al. Femtosecond laser: a new intradermal DNA delivery method for efficient, long-term gene expression and genetic immunization. *FASEB J.* 2007;21(13):3522-33.
[6] Cattamanchi A, Posavad CM, Wald A, et al. Phase I study of a herpes simplex virus type 2 (HSV-2) DNA vaccine administered to healthy, HSV-2-seronegative adults by a needle-free injection system. *Clin.Vaccine Immunol.* 2008;15(11):1638-43.
[7] Johnston KB, Monteiro JM, Schultz LD, et al. Protection of beagle dogs from mucosal challenge with canine oral papillomavirus by immunization with recombinant adenoviruses expressing codon-optimized early genes 3. Virology 2005;336(2):208-18.
[8] Donnelly JJ, Wahren B, Liu MA. DNA vaccines: progress and challenges. *J. Immunol.* 2005;175(2):633-9.
[9] Kim TW, Hung CF, Boyd D, et al. Enhancing DNA vaccine potency by combining a strategy to prolong dendritic cell life with intracellular targeting strategies. *J. Immunol.* 2003;171(6):2970-6.
[10] Lu S, Wang S, Grimes-Serrano JM. Current progress of DNA vaccine studies in humans. *Expert.Rev.Vaccines.* 2008;7(2):175-91.
[11] Kibuuka H, Kimutai R, Maboko L, et al. A phase 1/2 study of a multiclade HIV-1 DNA plasmid prime and recombinant adenovirus serotype 5 boost vaccine in HIV-Uninfected East Africans (RV 172). *J. Infect.Dis.* 2010;201(4):600-7.
[12] Smith LR, Wloch MK, Ye M, et al. Phase 1 clinical trials of the safety and immunogenicity of adjuvanted plasmid DNA vaccines encoding influenza A virus H5 hemagglutinin. *Vaccine* 2010;28(13):2565-72.
[13] Hokey DA, Weiner DB. DNA vaccines for HIV: challenges and opportunities. *Springer Semin. Immunopathol.* 2006;28(3):267-79.
[14] Liu MA. Overview of DNA vaccines. *Ann. NY Acad.Sci.* 1995;772:15-20.
[15] Donnelly JJ, Ulmer JB, Liu MA. DNA vaccines. *Life Sci.* 1997;60:163

[16] Wolff JA, Budker V. The mechanism of naked DNA uptake and expression. *Adv.Genet.* 2005;54:3-20.
[17] Kutzler MA, Weiner DB. Developing DNA vaccines that call to dendritic cells. *J. Clin. Invest* 2004;114(9):1241-4.
[18] Herweijer H, Zhang G, Subbotin VM, Budker V, Williams P, Wolff JA. Time course of gene expression after plasmid DNA gene transfer to the liver. *J. Gene Med.* 2001;3(3):280-91.
[19] Wolff JA, Ludtke JJ, Acsadi G, Williams P, Jani A. Long-term persistence of plasmid DNA and foreign gene expression in mouse muscle. *Hum. Mol. Genet.* 1992;1(6):363-9.
[20] Zhang L, Widera G, Bleecher S, Zaharoff DA, Mossop B, Rabussay D. Accelerated immune response to DNA vaccines. *DNA Cell Biol.* 2003;22(12):815-22.
[21] Brandsma JL, Shlyankevich M, Zhang L, et al. Vaccination of rabbits with an adenovirus vector expressing the papillomavirus E2 protein leads to clearance of papillomas and infection. *J.Virol.* 2004;78(1):116-23.
[22] Katsube K, Bishop AT, Friedrich PF. Transduction of rabbit saphenous artery: a comparison of naked DNA, liposome complexes, and adenovirus vectors. *J. Orthop. Res.* 2004;22(6):1290-5.
[23] Ingolotti M, Kawalekar O, Shedlock DJ, Muthumani K, Weiner DB. DNA vaccines for targeting bacterial infections. Expert.Rev.Vaccines. 2010;9(7):747-63.
[24] Luo D. A new solution for improving gene delivery. *Trends Biotechnol.* 2004;22(3):101-3.
[25] Marwick C. Exciting potential of DNA vaccines explored. *JAMA* 1995;273(18):1403
[26] Boyd D, Hung CF, Wu TC. DNA vaccines for cancer. *IDrugs.* 2003;6(12):1155-64.
[27] Ribas A, Butterfield LH, Glaspy JA, Economou JS. Current developments in cancer vaccines and cellular immunotherapy. *J.Clin.Oncol.* 2003;21(12):2415-32.
[28] Lin K, Roosinovich E, Ma B, Hung CF, Wu TC. Therapeutic HPV DNA vaccines. *Immunol. Res.* 2010;47(1-3):86-112.
[29] Ling M, Kanayama M, Roden R, Wu TC. Preventive and therapeutic vaccines for human papillomavirus-associated cervical cancers. *J. Biomed.Sci.* 2000;7(5):341-56.
[30] Christensen ND. Cottontail rabbit papillomavirus (CRPV) model system to test antiviral and immunotherapeutic strategies. *Antivir.Chem.Chemother.* 2005;16(6):355-62.
[31] Campo MS. Animal models of papillomavirus pathogenesis. *Virus Res.* 2002; 89(2): 249-61.
[32] Nicholls PK, Stanley MA. The immunology of animal papillomaviruses. *Vet. Immunol. Immunopathol.* 2000;73(2):101-27.
[33] Bolhassani A, Mohit E, Rafati S. Different spectra of therapeutic vaccine development against HPV infections. *Hum.Vaccin.* 2009;5(10)
[34] Stanley MA, Moore RA, Nicholls PK, et al. Intra-epithelial vaccination with COPV L1 DNA by particle-mediated DNA delivery protects against mucosal challenge with infectious COPV in beagle dogs. *Vaccine* 2001;19(20-22):2783-92.
[35] Moore RA, Walcott S, White KL, et al. Therapeutic immunisation with COPV early genes by epithelial DNA delivery 2. *Virology* 2003;314(2):630-5.
[36] Han R, Reed CA, Cladel NM, Christensen ND. Intramuscular injection of plasmid DNA encoding cottontail rabbit papillomavirus E1, E2, E6 and E7 induces T cell-

mediated but not humoral immune responses in rabbits. *Vaccine* 1999;17(11-12):1558-66.

[37] Leachman SA, Tigelaar RE, Shlyankevich M, et al. Granulocyte-macrophage colony-stimulating factor priming plus papillomavirus E6 DNA vaccination: effects on papilloma formation and regression in the cottontail rabbit papillomavirus--rabbit model. *J.Virol.* 2000;74(18):8700-8.

[38] Hu J, Han R, Cladel NM, Pickel MD, Christensen ND. Intracutaneous DNA vaccination with the E8 gene of cottontail rabbit papillomavirus induces protective immunity against virus challenge in rabbits. *J.Virol.* 2002;76(13):6453-9.

[39] Hu J, Cladel NM, Budgeon LR, Reed CA, Pickel MD, Christensen ND. Protective cell-mediated immunity by DNA vaccination against Papillomavirus L1 capsid protein in the Cottontail Rabbit Papillomavirus model. *Viral Immunol.* 2006;19(3):492-507.

[40] Sundaram P, Tigelaar RE, Xiao W, Brandsma JL. Intracutaneous vaccination of rabbits with the E6 gene of cottontail rabbit papillomavirus provides partial protection against virus challenge. *Vaccine* 1998;16(6):613-23.

[41] Brandsma JL, Shlyankevich M, Zelterman D, Su Y. Therapeutic vaccination of rabbits with a ubiquitin-fused papillomavirus E1, E2, E6 and E7 DNA vaccine. *Vaccine* 2007;25(33):6158-63.

[42] Leachman SA, Shylankevich M, Slade MD, et al. Ubiquitin-fused and/or multiple early genes from cottontail rabbit papillomavirus as DNA vaccines. *J.Virol.* 2002;76(15):7616-24.

[43] Hu J, Cladel NM, Christensen ND. Increased immunity to cottontail rabbit papillomavirus infection in EIII/JC inbred rabbits after vaccination with a mutant E6 that correlates with spontaneous regression. *Viral Immunol.* 2007;20(2):320-5.

[44] Brandsma JL, Shlyankevich M, Su Y, Zelterman D, Rose JK, Buonocore L. Reversal of papilloma growth in rabbits therapeutically vaccinated against E6 with naked DNA and/or vesicular stomatitis virus vectors. *Vaccine* 2009;

[45] Han R, Reed CA, Cladel NM, Christensen ND. Immunization of rabbits with cottontail rabbit papillomavirus E1 and E2 genes: protective immunity induced by gene gun-mediated intracutaneous delivery but not by intramuscular injection. *Vaccine* 2000;18(26):2937-44.

[46] Han R, Cladel NM, Reed CA, Peng X, Christensen ND. Protection of rabbits from viral challenge by gene gun-based intracutaneous vaccination with a combination of cottontail rabbit papillomavirus E1, E2, E6, and E7 genes. J.Virol. 1999;73(8):7039-43.

[47] Pink JRL, Sinigaglia F. Characterizing T-Cell Epitopes in Vaccine Candidates. *Immunol.Today* 1989;10:408-9.

[48] Sinigaglia F, Romagnoli P, Guttinger M, Takacs B, Pink JRL. Selection of T cell epitopes and vaccine engineering. *Methods Enzymol.* 1991;203:370-86.

[49] Wang QM, Sun SH, Hu ZL, et al. Epitope DNA vaccines against tuberculosis: spacers and ubiquitin modulates cellular immune responses elicited by epitope DNA vaccine. *Scand. J. Immunol.* 2004;60(3):219-25.

[50] Ma M, Jin N, Yin G, et al. [Molecular design and immunogenicity of a multiple-epitope foot-and-mouth disease virus antigen, adjuvants, and DNA vaccination]. *Sheng Wu Gong. Cheng Xue. Bao.* 2009;25(4):514-9.

[51] Steller MA, Gurski KJ, Murakami M, et al. Cell-mediated immunological responses in cervical and vaginal cancer patients immunized with a lipidated epitope of human papillomavirus type 16 E7. *Clin.Cancer Res.* 1998;4(9):2103-9.

[52] Ghosh MK, Li CL, Fayolle C, et al. Induction of HLA-A2-restricted CTL responses by a tubular structure carrying human melanoma epitopes. *Vaccine* 2002;20(19-20):2463-73.

[53] Velders MP, Weijzen S, Eiben GL, et al. Defined flanking spacers and enhanced proteolysis is essential for eradication of established tumors by an epitope string DNA vaccine. *J. Immunol.* 2001;166(9):5366-73.

[54] Thomson SA, Sherritt MA, Medveczky J, et al. Delivery of multiple CD8 cytotoxic T cell epitopes by DNA vaccination. *J. Immunol.* 1998;160(4):1717-23.

[55] Huber VC, Thomas PG, McCullers JA. A multi-valent vaccine approach that elicits broad immunity within an influenza subtype. *Vaccine* 2009;27(8):1192-200.

[56] Fynan EF, Robinson HL, Webster RG. Use of DNA encoding influenza hemagglutinin as an avian influenza vaccine. *DNA Cell Biol.* 1993;12:785-9.

[57] Chang H, Huang C, Wu J, et al. A single dose of DNA vaccine based on conserved H5N1 subtype proteins provides protection against lethal H5N1 challenge in mice pre-exposed to H1N1 influenza virus. *Virol. J.* 2010;7:197

[58] Choi SY, Suh YS, Cho JH, Jin HT, Chang J, Sung YC. Enhancement of DNA Vaccine-induced Immune Responses by Influenza Virus NP Gene. *Immune. Netw.* 2009;9(5):169-78.

[59] Ramachandran S, Kinchington PR. Potential prophylactic and therapeutic vaccines for HSV infections. *Curr. Pharm. Des.* 2007;13(19):1965-73.

[60] Jamali A, Roostaee MH, Soleimanjahi H, Ghaderi PF, Bamdad T. DNA vaccine-encoded glycoprotein B of HSV-1 fails to protect chronic morphine-treated mice against HSV-1 challenge. *Comp. Immunol.Microbiol.Infect. Dis.* 2007;30(2):71-80.

[61] BenMohamed L, Bertrand G, McNamara CD, et al. Identification of novel immunodominant CD4+ Th1-type T-cell peptide epitopes from herpes simplex virus glycoprotein D that confer protective immunity. *J.Virol.* 2003;77(17):9463-73.

[62] Chentoufi AA, Dasgupta G, Christensen ND, et al. A Novel HLA (HLA-A*0201) Transgenic Rabbit Model for Preclinical Evaluation of Human CD8+ T Cell Epitope-Based Vaccines against Ocular Herpes. *J. Immunol.* 2010;184(5):2561-71.

[63] Doorbar J. Molecular biology of human papillomavirus infection and cervical cancer. *Clin. Sci.(*Lond) 2006;110(5):525-41.

[64] Stanley MA. Progress in prophylactic and therapeutic vaccines for human papillomavirus infection 3. *Expert Rev.Vaccines.* 2003;2(3):381-9.

[65] Einstein MH, Baron M, Levin MJ, et al. Comparison of the immunogenicity and safety of Cervarix() and Gardasil((R)) human papillomavirus (HPV) cervical cancer vaccines in healthy women aged 18-45 years. *Hum.Vaccin.* 2009;5(10)

[66] Fausch SC, Da Silva DM, Eiben GL, Le Poole IC, Kast WM. HPV protein/peptide vaccines: from animal models to clinical trials. *Front Biosci.* 2003;8:s81-s91

[67] Rudolf MP, Nieland JD, DaSilva DM, et al. Induction of HPV16 capsid protein-specific human T cell responses by virus-like particles. *Biol.Chem.* 1999;380(3):335-40.

[68] Woodberry T, Gardner J, Mateo L, et al. Immunogenicity of a human immunodeficiency virus (HIV) polytope vaccine containing multiple HLA A2 HIV CD8(+) cytotoxic T-cell epitopes. *J.Virol.* 1999;73(7):5320-5.

[69] Qiu J, Luo P, Wasmund K, Steplewski Z, Kieber-Emmons T. Towards the development of peptide mimotopes of carbohydrate antigens as cancer vaccines. *Hybridoma* 1999;18(1):103-12.

[70] Jiang Y, Lin C, Yin B, et al. Effects of the configuration of a multi-epitope chimeric malaria DNA vaccine on its antigenicity to mice. *Chin. Med. J.(*Engl.) 1999; 112(8):686-90.

[71] Yang W, Jackson DC, Zeng Q, McManus DP. Multi-epitope schistosome vaccine candidates tested for protective immunogenicity in mice. *Vaccine* 2000;19(1):103-13.

[72] Sundaram R, Sun YP, Walker CM, Lemonnier FA, Jacobson S, Kaumaya PTP. A novel multivalent human CTL peptide construct elicits robust cellular immune responses in HLA-A*0201 transgenic mice: implications for HTLV-1 vaccine design. *Vaccine* 2003;21(21-22):2767-81.

[73] Sundaram R, Lynch MP, Rawale S, et al. Protective Efficacy of Multiepitope Human Leukocyte Antigen-A*0201 Restricted Cytotoxic T-Lymphocyte Peptide Construct Against Challenge With Human T-Cell Lymphotropic Virus Type 1 Tax Recombinant Vaccinia Virus. *J. Acquir. Immune. Defic.Syndr.* 2004;37(3):1329-39.

[74] Singh RA, Barry MA. Generation of multivalent genome-wide T cell responses in HLA-A*0201 transgenic mice by an HIV-1 expression library immunization (ELI) vaccine. *Res. Initiat.Treat. Action.* 2003;8(2):17-9.

[75] Hu J, Cladel N, Peng X, Balogh K, Christensen ND. Protective immunity with an E1 multivalent epitope DNA vaccine against cottontail rabbit papillomavirus (CRPV) infection in an HLA-A2.1 transgenic rabbit model 1. *Vaccine* 2008;26(6):809-16.

[76] Peachman KK, Rao M, Alving CR. Immunization with DNA through the skin. *Methods* 2003;31(3):232-42.

[77] Kutzler MA, Weiner DB. DNA vaccines: ready for prime time? Nat.Rev.Genet. 2008;9(10):776-88.

[78] Hu J, Peng X, Schell TD, Budgeon LR, Cladel NM, Christensen ND. An HLA-A2.1-Transgenic Rabbit Model to Study Immunity to Papillomavirus Infection. *J. Immunol.* 2006;177(11):8037-45.

[79] Galvin TA, Muller J, Khan AS. Effect of different promoters on immune responses elicited by HIV-1 gag/env multigenic DNA vaccine in macaca mulatta and macaca nemestrina [In Process Citation]. *Vaccine* 2000;18(23):2566-83.

[80] Belakova J, Horynova M, Krupka M, Weigl E, Raska M. DNA vaccines: are they still just a powerful tool for the future? *Arch. Immunol.Ther. Exp.(*Warsz.) 2007;55(6):387-98.

[81] Qin L, Ding Y, Pahud DR, Chang E, Imperiale MJ, Bromberg JS. Promoter attenuation in gene therapy: interferon-gamma and tumor necrosis factor-alpha inhibit transgene expression [see comments]. *Hum.Gene Ther.* 1997;8(17):2019-29.

[82] Lu D, Hoory T, Monie A, Wu A, Wang MC, Hung CF. Treatment with demethylating agent, 5-aza-2'-deoxycytidine enhances therapeutic HPV DNA vaccine potency. *Vaccine* 2009;27(32):4363-9.

[83] Zhou J, Liu WJ, Peng SW, Sun XY, Frazer I. Papillomavirus capsid protein expression level depends on the match between codon usage and tRNA availability. J.Virol. 1999;73(6):4972-82.

[84] Coleman JR, Papamichail D, Skiena S, Futcher B, Wimmer E, Mueller S. Virus attenuation by genome-scale changes in codon pair bias. *Science* 2008;320(5884):1784-7.

[85] Shackelton LA, Parrish CR, Holmes EC. Evolutionary basis of codon usage and nucleotide composition bias in vertebrate DNA viruses. *J. Mol. Evol.* 2006;62(5):551-63.

[86] Tindle RW. Immune evasion in human papillomavirus-associated cervical cancer. Nature Rev.*Cancer* 2002;2(1):59-65.

[87] Cid-Arregui A, Juarez V, zur HH. A synthetic E7 gene of human papillomavirus type 16 that yields enhanced expression of the protein in mammalian cells and is useful for DNA immunization studies. *J.Virol.* 2003;77(8):4928-37.

[88] Leder C, Kleinschmidt JA, Wiethe C, Muller M. Enhancement of capsid gene expression: preparing the human papillomavirus type 16 major structural gene L1 for DNA vaccination purposes. *J.Virol.* 2001;75(19):9201-9.

[89] Rollman E, Arnheim L, Collier B, et al. HPV-16 L1 genes with inactivated negative RNA elements induce potent immune responses. *Virology* 2004;322(1):182-9.

[90] Mossadegh N, Gissmann L, Muller M, Zentgraf H, Alonso A, Tomakidi P. Codon optimization of the human papillomavirus 11 (HPV 11) L1 gene leads to increased gene expression and formation of virus-like particles in mammalian epithelial cells 344. *Virology* 2004;326(1):57-66.

[91] Cladel NM, Hu J, Balogh KK, Christensen ND. CRPV genomes with synonymous codon optimizations in the CRPV E7 gene show phenotypic differences in growth and altered immunity upon E7 vaccination. *PLoS.One.* 2008;3(8):e2947

[92] Liu WJ, Gao F, Zhao KN, et al. Codon modified human papillomavirus type 16 E7 DNA vaccine enhances cytotoxic T-lymphocyte induction and anti-tumour activity. *Virology* 2002;301(1):43-52.

[93] Liu WJ, Zhao KN, Gao FG, Leggatt GR, Fernando GJ, Frazer IH. Polynucleotide viral vaccines: codon optimisation and ubiquitin conjugation enhances prophylactic and therapeutic efficacy. *Vaccine* 2001;20(5-6):862-9.

[94] Fomsgaard A. HIV-1 DNA vaccines. *Immunol. Lett.* 1999;65(1-2):127-31.

[95] Wu X, Jornvall H, Berndt KD, Oppermann U. Codon optimization reveals critical factors for high level expression of two rare codon genes in Escherichia coli: RNA stability and secondary structure but not tRNA abundance. *Biochem. Biophys. Res. Commun.* 2004;313(1):89-96.

[96] Dobano C, Sedegah M, Rogers WO, et al. Plasmodium: mammalian codon optimization of malaria plasmid DNA vaccines enhances antibody responses but not T cell responses nor protective immunity. *Exp. Parasitol.* 2009;122(2):112-23.

[97] Abdulhaqq SA, Weiner DB. DNA vaccines: developing new strategies to enhance immune responses. *Immunol. Res.* 2008;42(1-3):219-32.

[98] Scheerlinck JY. Genetic adjuvants for DNA vaccines. *Vaccine* 2001;19(17-19):2647-56.

[99] Hu J, Cladel NM, Wang Z, Han R, Pickel MD, Christensen ND. GM-CSF enhances protective immunity to cottontail rabbit papillomavirus E8 genetic vaccination in rabbits. *Vaccine* 2004;22(9-10):1124-30.

[100] Gurunathan S, Sacks DL, Brown DR, et al. Vaccination with DNA encoding the immunodominant LACK parasite antigen confers protective immunity to mice infected with Leishmania major. *J. Exp. Med.* 1997;186(7):1137-47.

[101] Klinman DM, Klaschik S, Sato T, Tross D. CpG oligonucleotides as adjuvants for vaccines targeting infectious diseases. *Adv. Drug Deliv. Rev.* 2009;61(3):248-55.

[102] Coban C, Ishii KJ, Gursel M, Klinman DM, Kumar N. Effect of plasmid backbone modification by different human CpG motifs on the immunogenicity of DNA vaccine vectors. *J. Leukoc. Biol.* 2005;78(3):647-55.

[103] Daftarian P, Ali S, Sharan R, et al. Immunization with Th-CTL fusion peptide and cytosine-phosphate-guanine DNA in transgenic HLA-A2 mice induces recognition of HIV-infected T cells and clears vaccinia virus challenge. *J. Immunol.* 2003;171(8):4028-39.

[104] Grossmann C, Tenbusch M, Nchinda G, et al. Enhancement of the priming efficacy of DNA vaccines encoding dendritic cell-targeted antigens by synergistic toll-like receptor ligands. *BMC. Immunol.* 2009;10:43

[105] Alpar HO, Papanicolaou I, Bramwell VW. Strategies for DNA vaccine delivery. *Expert.Opin. Drug Deliv.* 2005;2(5):829-42.

[106] Keegan ME, Saltzman WM. Surface-modified biodegradable microspheres for DNA vaccine delivery. *Methods Mol. Med.* 2006;127:107-13.

[107] Smorlesi A, Papalini F, Amici A, et al. Evaluation of different plasmid DNA delivery systems for immunization against HER2/neu in a transgenic murine model of mammary carcinoma. *Vaccine* 2006;24(11):1766-75.

[108] Lisziewicz J, Calarota SA, Lori F. The potential of topical DNA vaccines adjuvanted by cytokines. *Expert.Opin. Biol.Ther.* 2007;7(10):1563-74.

[109] Luo D, Saltzman WM. Synthetic DNA delivery systems. *Nat. Biotechnol.* 2000;18(1):33-7.

[110] Donnelly JJ, Wahren B, Liu MA. DNA vaccines: progress and challenges. *J. Immunol.* 2005;175(2):633-9.

[111] Choi MJ, Kim JH, Maibach HI. Topical DNA vaccination with DNA/Lipid based complex. *Curr. Drug Deliv.* 2006;3(1):37-45.

[112] Rosenberg ES, Graham BS, Chan ES, et al. Safety and immunogenicity of therapeutic DNA vaccination in individuals treated with antiretroviral therapy during acute/early HIV-1 infection. *PLoS.One.* 2010;5(5):e10555

[113] Widera G, Austin M, Rabussay D, et al. Increased DNA vaccine delivery and immunogenicity by electroporation in vivo. J.Immunol. 2000;164(9):4635-40.

[114] Spencer SC. Electroporation technique of DNA transfection. *Appl. Biochem. Biotechnol.* 1993;42:75-82.

[115] Rols MP. Mechanism by which electroporation mediates DNA migration and entry into cells and targeted tissues. *Methods Mol. Biol.* 2008;423:19-33.

[116] Babiuk LA, Pontarollo R, Babiuk S, Loehr B, van Drunen Littel-van den Hurk. Induction of immune responses by DNA vaccines in large animals. *Vaccine* 2003;21(7-8):649-58.

[117] Luxembourg A, Evans CF, Hannaman D. Electroporation-based DNA immunisation: translation to the clinic. Expert.Opin.Biol.Ther. 2007;7(11):1647-64.

[118] Anwer K. Formulations for DNA delivery via electroporation in vivo. *Methods Mol.Biol.* 2008;423:77-89.

[119] Bertling W, Hunger-Bertling K, Cline MJ. Intranuclear uptake and persistence of biologically active DNA after electroporation of mammalian cells. *J. Biochem. Biophys. Methods* 1987;14:223-32.

[120] Pokorna D, Polakova I, Kindlova M, et al. Vaccination with human papillomavirus type 16-derived peptides using a tattoo device. *Vaccine* 2009;27(27):3519-29.

[121] van den Berg JH, Nujien B, Beijnen JH, et al. Optimization of intradermal vaccination by DNA tattooing in human skin. *Hum.Gene Ther.* 2009;20(3):181-9.

[122] Bins AD, Jorritsma A, Wolkers MC, et al. A rapid and potent DNA vaccination strategy defined by in vivo monitoring of antigen expression. *Nat. Med.* 2005;11(8):899-904.

[123] Pokorna D, Rubio I, Muller M. DNA-vaccination via tattooing induces stronger humoral and cellular immune responses than intramuscular delivery supported by molecular adjuvants. *Genet.Vaccines.Ther.* 2008;6:4

[124] Quaak SG, van den Berg JH, Oosterhuis K, Beijnen JH, Haanen JB, Nuijen B. DNA tattoo vaccination: effect on plasmid purity and transfection efficiency of different topoisoforms. *J.Control Release* 2009;139(2):153-9.

[125] Streilein JW. Circuits and signals of the skin-associated lymphoid tissues (SALT). *J. Invest. Dermatol.* 1985;85:10s-3s.

[126] Bos JD, Kapsenberg ML. The skin immune system: progress in cutaneous biology. *Immunol.Today* 1993;14:75-8.

[127] Memar OM, Arany I, Tyring SK. Skin-associated lymphoid tissue in human immunodeficiency virus-1, human papillomavirus, and herpes simplex virus infections. *J. Invest. Dermatol.* 1995;105 Suppl.:99S-104S.

[128] Ozden S, Cochet M, Mikol J, Teixeira A, Gessain A, Pique C. Direct evidence for a chronic CD8+-T-cell-mediated immune reaction to tax within the muscle of a human T-cell leukemia/lymphoma virus type 1-infected patient with sporadic inclusion body myositis. *J.Virol.* 2004;78(19):10320-7.

[129] Daugimont L, Baron N, Vandermeulen G, et al. Hollow microneedle arrays for intradermal drug delivery and DNA electroporation. *J. Membr. Biol.* 2010;236(1):117-25.

[130] Syed TA, Khayyami M, Kriz D, Svanberg K, Kahlon RC, Ahmad SA. Management of genital warts in women with human leukocyte interferon-□ vs. podophyllotoxin in cream: A placebo-controlled, double-blind, comparative study. *J. Mol. Med.* 1995;73:255-8.

[131] Prow TW, Chen X, Prow NA, et al. Nanopatch-Targeted Skin Vaccination against West Nile Virus and Chikungunya Virus in Mice. *Small* 2010;6(16):1776-84.

[132] Kask AS, Chen X, Marshak JO, et al. DNA vaccine delivery by densely-packed and short microprojection arrays to skin protects against vaginal HSV-2 challenge. *Vaccine* 2010;28(47):7483-91.

[133] Chen X, Kask AS, Crichton ML, et al. Improved DNA vaccination by skin-targeted delivery using dry-coated densely-packed microprojection arrays. *J. Control Release* 2010;148(3):327-33.

[134] Fynan EF, Webster RG, Fuller DH, Haynes JR, Santoro JC, Robinson HL. DNA vaccines -protective immunizations by parenteral, mucosal, and gene-gun inoculations. *Proc. Natl. Acad.Sci.USA* 1993;90(24):11478

[135] Pertmer TM, Eisenbraun MD, McCabe D, Prayaga SK, Fuller DH, Haynes JR. Gene gun-based nucleic acid immunization: elicitation of humoral and cytotoxic T

lymphocyte responses following epidermal delivery of nanogram quantities of DNA. *Vaccine* 1995;13(15):1427-30.
[136] Haynes JR, McCabe DE, Swain WF, Widera G, Fuller JT. Particle-mediated nucleic acid immunization. *J. Biotechnol.* 1996;44(1-3):37-42.
[137] Mumper RJ, Ledebur HC, Jr. Dendritic cell delivery of plasmid DNA. Applications for controlled genetic immunization. *Mol. Biotechnol.* 2001;19(1):79-95.
[138] Hung CF, Monie A, Alvarez RD, Wu TC. DNA vaccines for cervical cancer: from bench to bedside. *Exp. Mol. Med.* 2007;39(6):679-89.
[139] Wang S, Zhang C, Zhang L, Li J, Huang Z, Lu S. The relative immunogenicity of DNA vaccines delivered by the intramuscular needle injection, electroporation and gene gun methods. *Vaccine* 2008;26(17):2100-10.
[140] Yager EJ, Dean HJ, Fuller DH. Prospects for developing an effective particle-mediated DNA vaccine against influenza. Expert.Rev.*Vaccines.* 2009;8(9):1205-20.
[141] Trimble C, Lin CT, Hung CF, et al. Comparison of the CD8+ T cell responses and antitumor effects generated by DNA vaccine administered through gene gun, biojector, and syringe. *Vaccine* 2003;21(25-26):4036-42.
[142] Huang CF, Monie A, Weng WH, Wu T. DNA vaccines for cervical cancer. *Am. J. Transl. Res.* 2010;2(1):75-87.
[143] Lin K, Roosinovich E, Ma B, Hung CF, Wu TC. Therapeutic HPV DNA vaccines. *Immunol. Res.* 2010;47(1-3):86-112.
[144] Han R, Cladel NM, Reed CA, et al. DNA vaccination prevents and/or delays carcinoma development of papillomavirus-induced skin papillomas on rabbits. *J.Virol.* 2000;74:9712-6.
[145] Hu J, Schell T, Peng, X, Cladel NM, Balogh KK and Christensen ND. Strong and Specific Protective and Therapeutic Immunity Induced by Single HLA-A2.1 Restricted Epitope DNA Vaccine in Rabbits. , 4-14. 1-1-2009.
[146] Huang HN, Li TL, Chan YL, Chen CL, Wu CJ. Transdermal immunization with low-pressure-gene-gun mediated chitosan-based DNA vaccines against Japanese encephalitis virus. *Biomaterials* 2009;30(30):6017-25.
[147] Brave A, Ljungberg K, Wahren B, Liu MA. Vaccine delivery methods using viral vectors. *Mol. Pharm.* 2007;4(1):18-32.
[148] de Jonge J, Leenhouts JM, Holtrop M, et al. Cellular gene transfer mediated by influenza virosomes with encapsulated plasmid DNA. Biochem.J 2007;405(1):41-9.
[149] Carrasco L. Entry of animal viruses and macromolecules into cells. *FEBS Lett.* 1994;350(2-3):151-4.
[150] Touze A, Coursaget P. In vitro gene transfer using human papillomavirus-like particles. *Nucleic Acids Res.* 1998;26(5):1317-23.
[151] Rose RC, Bonnez W, Reichman RC, Garcea RL. Expression of human papillomavirus type 11 L1 protein in insect cells: in vivo and in vitro assembly of viruslike particles. *J.Virol.* 1993;67:1936-44.
[152] Zhou J, Sun XY, Stenzel DJ, Frazer IH. Expression of vaccinia recombinant HPV-16, L1 and L2 ORF proteins in epithelial cells is sufficient for assembly of HPV virion-like particles. *Virology* 1991;185:251-7.
[153] Buck CB, Pastrana DV, Lowy DR, Schiller JT. Generation of HPV pseudovirions using transfection and their use in neutralization assays. *Methods Mol. Med.* 2005;119:445-62.

[154] Malboeuf CM, Simon DA, Lee YE, et al. Human papillomavirus-like particles mediate functional delivery of plasmid DNA to antigen presenting cells in vivo. *Vaccine* 2007;25(17):3270-6.

[155] Chailertvanitkul VA, Pouton CW. Adenovirus: a blueprint for non-viral gene delivery. *Curr. Opin. Biotechnol.* 2010;21(5):627-32.

[156] Cui Z, Mumper RJ. Microparticles and nanoparticles as delivery systems for DNA vaccines. *Crit. Rev.Ther.Drug Carrier Syst.* 2003;20(2-3):103-37.

[157] Shahiwala A, Vyas TK, Amiji MM. Nanocarriers for systemic and mucosal vaccine delivery. *Recent Pat Drug Deliv. Formul.* 2007;1(1):1-9.

[158] Luten J, van Nostrum CF, De Smedt SC, Hennink WE. Biodegradable polymers as non-viral carriers for plasmid DNA delivery. *J. Control Release* 2008;126(2):97-110.

[159] Pozo-Rodriguez A, Delgado D, Solinis MA, et al. Solid lipid nanoparticles as potential tools for gene therapy: in vivo protein expression after intravenous administration. *Int. J. Pharm.* 2010;385(1-2):157-62.

[160] Ohagan DT, Rahman D, Mcgee JP, et al. Biodegradable microparticles as controlled release antigen delivery systems. *Immunology* 1991;73:239-42.

[161] Bertling WM, Gareis M, Paspaleeva V, et al. Use of liposomes, viral capsids and nanoparticles as DNA carriers. *Biotechnol. Appl. Biochem.* 1991;13:390-405.

[162] Patil SD, Rhodes DG, Burgess DJ. DNA-based therapeutics and DNA delivery systems: a comprehensive review. *AAPS. J.* 2005;7(1):E61-E77

[163] Storrie H, Mooney DJ. Sustained delivery of plasmid DNA from polymeric scaffolds for tissue engineering. *Adv. Drug Deliv. Rev.* 2006;58(4):500-14.

[164] Wang C-Y, Huang L. Plasmid DNA adsorbed to pH-sensitive liposomes efficiently transforms the target cells. *Biochem. Biophys. Res. Commun.* 1987;147:980-5.

[165] Christensen D, Agger EM, Andreasen LV, Kirby D, Andersen P, Perrie Y. Liposome-based cationic adjuvant formulations (CAF): past, present, and future. *J. Liposome Res.* 2009;19(1):2-11.

[166] Tsen SW, Wu CY, Meneshian A, Pai SI, Hung CF, Wu TC. Femtosecond laser treatment enhances DNA transfection efficiency in vivo. *J. Biomed.Sci.* 2009;16:36

[167] Tirlapur UK, Konig K. Femtosecond near-infrared laser pulses as a versatile non-invasive tool for intra-tissue nanoprocessing in plants without compromising viability. *Plant J.* 2002;31(3):365-74.

[168] Chakravarty P, Qian W, El Sayed MA, Prausnitz MR. Delivery of molecules into cells using carbon nanoparticles activated by femtosecond laser pulses. *Nat. Nanotechnol.* 2010;5(8):607-11.

[169] Bao S, Thrall BD, Miller DL. Transfection of a reporter plasmid into cultured cells by sonoporation in vitro. *Ultrasound Med. Biol.* 1997;23(6):953-9.

[170] Fernandez-Alonso M, Rocha A, Coll JM. DNA vaccination by immersion and ultrasound to trout viral haemorrhagic septicaemia virus. *Vaccine* 2001;19(23-24):3067-75.

[171] Navot N, Kimmel E, Avtalion RR. Immunisation of fish by bath immersion using ultrasound. *Dev.Biol.(*Basel) 2005;121:135-42.

[172] Yoon CS, Park JH. Ultrasound-mediated gene delivery. *Expert.Opin. Drug Deliv.* 2010;7(3):321-30.

[173] Guo H, Leung JC, Chan LY, et al. Ultrasound-contrast agent mediated naked gene delivery in the peritoneal cavity of adult rat. *Gene Ther.* 2007;14(24):1712-20.

[174] Postema M, Gilja OH. Ultrasound-directed drug delivery. *Curr.Pharm.Biotechnol.* 2007;8(6):355-61.

[175] Zarnitsyn VG, Kamaev PP, Prausnitz MR. Ultrasound-enhanced chemotherapy and gene delivery for glioma cells. *Technol.Cancer Res.Treat.* 2007;6(5):433-42.

[176] Park EJ, Werner J, Smith NB. Ultrasound mediated transdermal insulin delivery in pigs using a lightweight transducer. *Pharm. Res.* 2007;24(7):1396-401.

[177] Wells DJ. Electroporation and ultrasound enhanced non-viral gene delivery in vitro and in vivo. *Cell Biol.Toxicol. 2010*;26(1):21-8.

[178] Smith NB, Lee S, Maione E, Roy RB, McElligott S, Shung KK. Ultrasound-mediated transdermal transport of insulin in vitro through human skin using novel transducer designs. *Ultrasound Med. Biol.* 2003;29(2):311-7.

[179] Zhou XY, Liao Q, Pu YM, et al. Ultrasound-mediated microbubble delivery of pigment epithelium-derived factor gene into retina inhibits choroidal neovascularization. Chin *Med.J.* (Engl.) 2009;122(22):2711-7.

[180] Prausnitz MR. Microneedles for transdermal drug delivery. *Adv. Drug Deliv. Rev.* 2004;56(5):581-7.

[181] Kinoshita M, Hynynen K. A novel method for the intracellular delivery of siRNA using microbubble-enhanced focused ultrasound. *Biochem. Biophys. Res. Commun.* 2005;335(2):393-9.

[182] Kobulnik J, Kuliszewski MA, Stewart DJ, Lindner JR, Leong-Poi H. Comparison of gene delivery techniques for therapeutic angiogenesis ultrasound-mediated destruction of carrier microbubbles versus direct intramuscular injection. *J. Am. Coll. Cardiol.* 2009;54(18):1735-42.

[183] Luis J, Park EJ, Meyer RJ, Smith NB. Rectangular cymbal arrays for improved ultrasonic transdermal insulin delivery. *J. Acoust.Soc. Am.* 2007;122(4):2022-30.

[184] Luo D, Saltzman WM. Synthetic DNA delivery systems. *Nat. Biotechnol.* 2000;18(1):33-7.

[185] Rocha-Zavaleta L, Alejandre JE, Garcia-Carranca A. Parenteral and oral immunization with a plasmid DNA expressing the human papillomavirus 16-L1 gene induces systemic and mucosal antibodies and cytotoxic T lymphocyte responses. *J. Med.Virol.* 2002;66(1):86-95.

[186] Vandermeulen G, Daugimont L, Richiardi H, et al. Effect of tape stripping and adjuvants on immune response after intradermal DNA electroporation. *Pharm. Res.* 2009;26(7):1745-51.

[187] Hu J, Cladel N, Balogh K, Christensen N. Mucosally delivered peptides prime strong immunity in HLA-A2.1 transgenic rabbits. *Vaccine* 28 (2010) 3706–3713

[188] Wang S, Pal R, Mascola JR, et al. Polyvalent HIV-1 Env vaccine formulations delivered by the DNA priming plus protein boosting approach are effective in generating neutralizing antibodies against primary human immunodeficiency virus type 1 isolates from subtypes A, B, C, D and E. *Virology* 2006;350(1):34-47.

[189] Spaan WJM. Background paper: Progress towards a coronavirus recombinant DNA vaccine. *Adv. Exp. Med. Biol.* 1990;276:201-3.

[190] Brice GT, Dobano C, Sedegah M, et al. Extended immunization intervals enhance the immunogenicity and protective efficacy of plasmid DNA vaccines. *Microbes. Infect.* 2007;9(12-13):1439-46.

[191] Peng S, Trimble C, Alvarez RD, et al. Cluster intradermal DNA vaccination rapidly induces E7-specific CD8+ T-cell immune responses leading to therapeutic antitumor effects. *Gene Ther.* 2008;15(16):1156-66.

[192] Wurtele H, Little KC, Chartrand P. Illegitimate DNA integration in mammalian cells. *Gene Ther.* 2003;10(21):1791-9.

[193] Nichols WW, Ledwith BJ, Manam SV, Troilo PJ. Potential DNA vaccine integration into host cell genome. *Ann.NY Acad.Sci.* 1995;772:30-9.

[194] Robertson JS, Griffiths E. Assuring the quality, safety, and efficacy of DNA vaccines. *Methods Mol.Med.* 2006;127:363-74.

[195] Kang KK, Choi SM, Choi JH, et al. Safety evaluation of GX-12, a new HIV therapeutic vaccine: investigation of integration into the host genome and expression in the reproductive organs. *Intervirology* 2003;46(5):270-6.

[196] Ledwith BJ, Manam S, Troilo PJ, et al. Plasmid DNA vaccines: investigation of integration into host cellular DNA following intramuscular injection in mice. *Intervirology* 2000;43(4-6):258-72.

[197] Scheiblhofer S, Weiss R, Gabler M, Leitner WW, Thalhamer J. Replicase-based DNA vaccines for allergy treatment. *Methods Mol.Med.* 2006;127:221-35.

[198] Brudnak M, Miller KS. Expression cloning exploiting PCR rescue of transfected genes. *BioTechniques* 1993;14:66-8.

[199] Schneider JC, Jenings AF, Mun DM, McGovern PM, Chew LC. Auxotrophic markers pyrF and proC can replace antibiotic markers on protein production plasmids in high-cell-density Pseudomonas fluorescens fermentation. *Biotechnol.Prog.* 2005;21(2): 343-8.

In: DNA Vaccines: Types, Advantages and Limitations
Editors: E. C. Donnelly and A. M. Dixon
ISBN 978-1-61324-444-9


*Chapter IV*

# Recent Advances in DNA Vaccines

***Ieda Maria Longo Maugéri and Daniela Santoro Rosa***
Division of Immunology, Federal University of São Paulo UNIFESP
Rua Botucatu 862 4th floor, 04023-900, São Paulo, SP, Brazil

## Abstract

Traditional vaccine approaches have proven to be insufficient against a variety of diseases, mainly those without spontaneous cure. In the last years, recombinant DNA technology has emerged as a promising tool for vaccine development against infectious agents, cancer, autoimmunity and allergy. DNA vaccines are based on the delivery of genes encoding the antigen of interest that can be translated by host cells. Importantly, both humoral and cell-mediated immune responses may be elicited against multiple defined antigens simultaneously. It offers several potential advantages over conventional approaches, including safety profile and feasible production method. Despite these advantages, the limited immunogenicity of DNA vaccines in humans has hampered their use in the clinical setting. Recently, four DNA vaccines have been approved for veterinary use, suggesting in a near future their employment in humans. Further optimization of DNA vaccine technology, including rational plasmid design, different delivery systems, addition of adjuvants, mainly biological modifiers like some bacteria or their compounds, and improved immunization protocols is an interesting alternative to achieve the immunogenic properties that have been demonstrated in preclinical models. In this chapter we will discuss/focus on the main features of DNA vaccines.

## Introduction

The vaccinology field has evolved since Jenner´s first rational immunization experiment that let to discovery of smallpox vaccine. Since then, traditional ways of developing vaccines (such as inactivated or live attenuated virus vaccines) against infectious diseases have been highly successful for a number of pathogens. Despite these advances, development of effective vaccines against some major killers such as human immunodeficiency virus (HIV), malaria and tuberculosis have proved challenging.

In the early 1990s recombinant DNA vaccines have emerged as promising tools for vaccine development against infectious agents, cancer, autoimmunity and allergy [1]. DNA vaccines are based on the delivery of genes encoding for a specific protein antigen that can be transcribed and translated by host cells [2, 3]. Antigens can be presented by DNA in a suitable molecular form, ranging from full-length sequence to short MHC class I-or II-binding epitopes to optimize induction of T-cell responses [4]. This vaccination technology has shown promising results in eliciting both humoral and cell-mediated immune responses (including cytotoxic responses) and in generating protection against various pathogen challenges in preclinical models [1,5]. Furthermore, DNA vaccines have been evaluated as a potential therapy against infectious diseases, cancer, allergy and autoimmunity.

DNA vaccines offer several potential advantages over conventional approaches, including safety profile and feasible production method. However, in spite of high expectations based on their efficacy in preclinical models, DNA immunization showed less success in humans and to date no DNA vaccines have been licensed for human use. Immunogenicity of DNA vaccines varies significantly due to many factors including the inherent immunogenicity of the protein antigen encoded in the DNA vaccine, the optimal immune responses that can be achieved in different animal models and in humans with different genetic backgrounds and, to a great degree, the delivery methods used to administer the DNA vaccines [6]. Therefore, in recent years technical improvements have been developed to enhance its immunogenicity including gene optimization strategies, novel formulations and immune adjuvants, and more effective delivery approaches.

## DNA Vaccines X Traditional Vaccine Approaches

Traditional vaccine approaches include inactivated pathogens (influenza vaccine), live attenuated (polio, smallpox, yellow fever vaccines), purified proteins (Hepatitis B vaccine) or virus-like particles (Humam Papiloma Virus vaccine). The ability of such vaccines to induce protection is primarily based on antibody-dependent mechanisms. T cell responses to existing vaccines play also an important role but until recently they were not extensively analyzed. The development of effective vaccines against some viruses, bacteria and many parasites has been hampered by the fact that the humoral immune response does not seem to be the best effector arm of the immune system to provide protection [7]. Most of these are chronic diseases for which it is thought that strong cellular immunity, in particular cytotoxic T cells, is necessary to eliminate the cells that are infected with the pathogen.

New tools for vaccine modalities, including DNA vaccines, are being developed that could generate the appropriate immune response (humoral and/or cellular immunity). In addition, protection upon viral infection and cancer requires a cellular immune response mediated by cytotoxic T lymphocytes (CTLs). Recombinant DNA vaccines are a promising strategy to achieve these goals.

The observation that a protein-coding gene is able to elicit a specific immune response *in vivo* was first demonstrated by Tang and Johnston [8] who showed that delivering the human growth hormone gene directly into the skin of mice could elicit antigen-specific antibody responses. Since then numerous experiments were performed and showed that a vector encoding an organism protein injected *in vivo* offered a simple method to induce protective immune responses [9].

DNA vaccination might provide several important advantages over current vaccine approaches [10]. First, DNA vaccines are safer than live attenuated vaccines or inactivated viral vaccines since it is neither infectious nor capable of replication. Therefore, DNA vaccines are considered the safest vaccine platform available (Figure 1). This is due to the fact that, in general, the more likely a vaccine concept is to prove efficacious, the greater the safety concerns.

DNA vaccines can elicit both humoral and cellular immune responses against multiple defined antigens simultaneously and do not induce vector immunity in the host. Indeed, they can mimic the effects of live attenuated vaccines in their ability to induce major histocompatibility complex (MHC) class I restricted CD8 T-cell responses. In addition, DNA plasmids can be designed and manufactured in a relatively simple and cost-effective manner. Furthermore, they have high stability and relative temperature insensitivity that facilitates storage and shipping when compared to other vaccine modalities.

# Components of a DNA Vaccine

DNA vaccines or "naked DNA" are composed of foreign genetic material, in particular the genes that code for important antigens, cloned into an expression vector. An ideal vector for DNA vaccines should be safe in humans and easily produced at commercial scale. Expression plasmids used in DNA-based vaccination normally contain two units that are highlighted in Figure 2: (i) the *antigen expression unit* composed of: promoter/enhancer sequences (for optimal expression in mammalian cells), followed by antigen-encoding and polyadenylation sequences (for stabilization of mRNA transcripts) and (ii) the *production unit* composed of: bacterial sequences necessary for plasmid amplification and selection (i.e, an origin of replication allowing for growth in bacteria and a bacterial antibiotic resistance gene that allows for plasmid selection during bacterial culture [11]. A strong promoter may be required for optimal expression in mammalian cells. For this, some promoters derived from viruses such as cytomegalovirus (CMV) or simian virus 40 (SV40) have been used. The most commonly used selectable markers are bacterial antibiotic resistance genes, such as the ampicillin resistance gene. However, since the ampicillin resistance gene is precluded for use in humans, a kanamycin resistance gene is often used. Finally, the *Escherichia coli* ColE1 origin of replication, which is found in plasmids such as those in the pUC series, is most often used in DNA vaccines because it provides high plasmid copy numbers in bacteria enabling high yields of plasmid DNA upon purification.

In addition, DNA vaccines may also contain unmethylated cytosine-phosphate-guanine oligonucleotide sequences (CpG motifs), within the noncoding portion of the plasmid DNA. These unmethylated regions of DNA are common in bacterial genomes and may be found in the antibacterial resistance genes placed within the DNA plasmid. CpG motifs stimulate Toll-like receptor 9 (TLR9) [12] and trigger a cascade of activation, proliferation and differentiation of TLR9-expressing cells (antigen presenting cells). The activation of TLR9 results in the augmentation of the immune response against the antigen encoded by the plasmid in the expressing cell.

The construction of bacterial plasmids with vaccine inserts is accomplished using recombinant DNA technology. Once constructed, the vaccine plasmid is transformed into bacteria, where bacterial growth produces multiple plasmid copies. The plasmid DNA is then purified from the bacteria, by separating the circular plasmid from the much larger bacterial DNA and other bacterial impurities. This purified DNA acts as the vaccine.

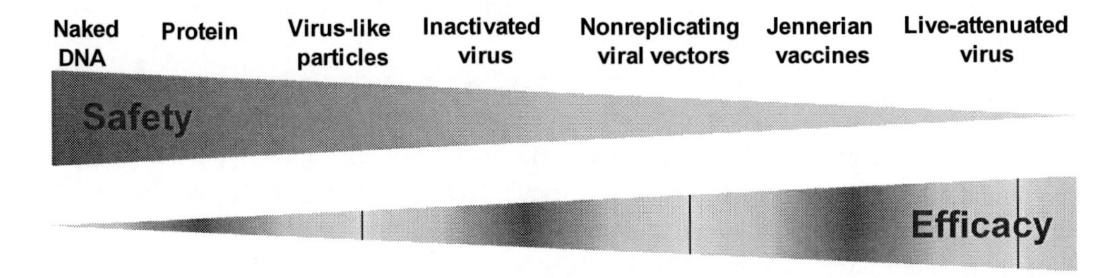


Figure 1. Safety and efficacy of different vaccine platforms.

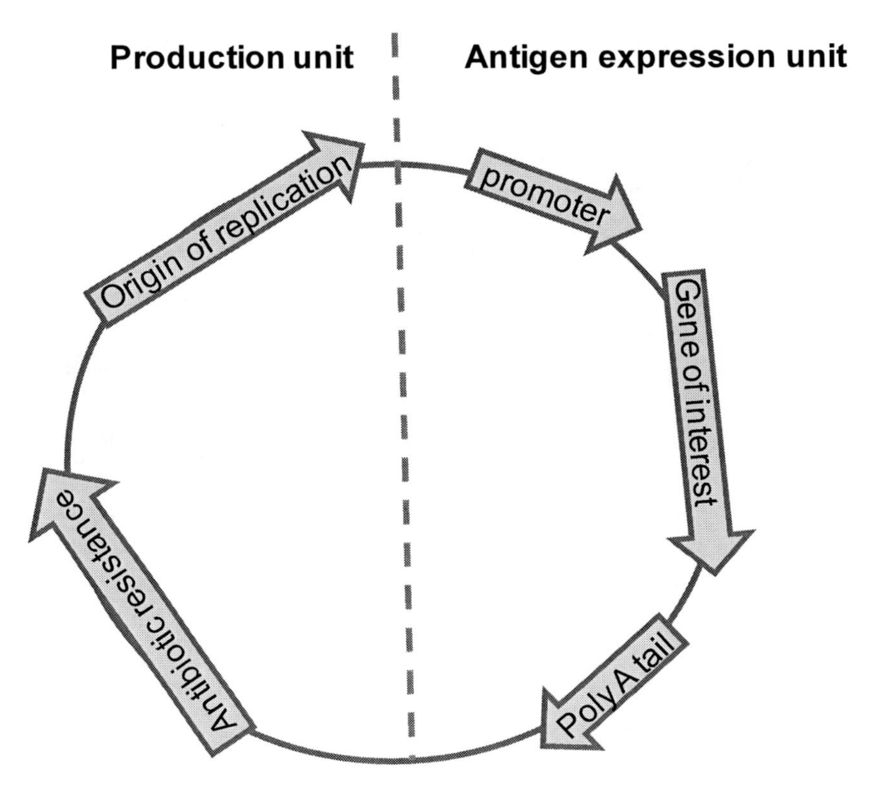


Figure 2. Schematic representation of a plasmid expression vector. The essential components of an expression vector include an antigen expression unit, which consists of a promoter, an insert containing the antigen and the Poly A tail. The other component is the production unit which contains bacterial sequences necessary for plasmid amplification and selection as a bacterial origin of replication and antibiotic resistance gene.

# How DNA Vaccines Work

For most viral and bacterial infections, primary protection is mediated by a humoral immune response. In the case of intracellular infections such as Mycobacterium *tuberculosis* and other parasites, cellular immune responses are needed. But beyond this, for some diseases like human immunodeficiency virus infection (HIV) and malaria both humoral and cellular responses are required.

After *in vivo* administration, DNA vaccines induce strong immune responses to the antigen encoded by the gene vaccine. Following DNA immunization, the plasmid enters the nucleus of transfected cells, initiates gene transcription and produces the corresponding protein inside the cell [13]. The form of antigen processing and presentation is able to induce both humoral and MHC class I and II restricted cellular immune responses. Secreted or exogenous proteins undergo endocytosis or phagocytosis to enter the MHC class II pathway of antigen processing to stimulate $CD4^+$ T cells. Endogenously produced proteins/peptides are presented to the immune system through an MHC class I dependent pathway to stimulate $CD8^+$ T cells.

Three principal mechanisms of antigen presentation after DNA vaccination are proposed: i) transfection of somatic cell (e.g. myocytes, keratinocytes) ; ii) transfection of professional antigen presenting cell (APC; e.g. dendritic cells); iii) uptake of secreted antigen and presentation by professional APCs through cross-priming.

*Transfection of somatic cells.* When somatic cells like myocytes and keratinocytes are transfected with DNA, the produced antigen is processed by the proteassome and the resulting peptides presented via class I MHC to T cells. Although muscle cells express class I molecules, they do not express costimulatory molecules (e.g. CD80 and CD86) that are critical to stimulate T-cell priming. Therefore, this cell type is not efficient in priming T cell responses as professional antigen presenting cells (e.g.dendritic cells). For other somatic cells like keratinocytes, it has been shown that they constitute one of the major cell types transfected by plasmid DNA after injection into the skin.

*Transfection of professional antigen presenting cells.* Direct transfection of APCs seems to be the most efficient method of priming a T cell response and dendritic cells are thought to play a key role. After antigen production, the endogenously synthesized protein is processed by the proteassome and the resulting peptides are presented via class I MHC to T cells. DNA uptake and gene expression have been observed in dendritic cells *in vivo* following DNA immunization [14] and adoptive transfer of the *in vivo* transfected cells leads to cytotoxic T cells (CTL) induction [15].

*Uptake of secreted antigen and presentation by professional APCs through cross-priming.* Alternatively, antigens synthesized after DNA vaccination can also be released from the transfected cells (e.g. somatic cells) into the extracellular milieu, and these soluble materials are taken up by specialized APCs which express both classes of MHC molecules [1]. Inside these APCs, the antigens enter the MHC class II pathway, and induce MHC class II restricted $CD4^+$ T cells, which usually secrete cytokines and provide "help" for B and $CD8^+$ T cells. Another mechanism that has been demonstrated is cross-priming: the transfected muscle cell produces the protein antigen, but then the antigen in some form is transferred to the APCs that are then directly responsible for activating the CTL responses [16]. Cross-priming can also occur when the transfected somatic cell undergoes apoptosis/necrosis and is engulfed by an APC [17]. Furthermore, soluble antigens can encounter B lymphocytes, be captured by specific high affinity immunoglobulins and therefore (in concert with CD4 T cell "help") induce an effective antibody response.

Independently of which mechanism the APCs are activated (by direct transfection or through cross-priming), antigen-loaded APCs migrate to the draining lymph nodes (DLN) via the afferent lymphatic vessel where they present peptide antigens to naive T cells (CD4 and CD8) via the interaction of the MHC and the T cell receptor (TCR) in combination with co-stimulatory molecules (e.g. CD80 and CD86). Once activated, these T cells expand and migrate through the efferent lymphatic vessel.

## Delivery Methods

The DNA site and delivery method may affect the nature of the induced immune response against antigens encoded by the plasmid [18, 19]. Previous studies showed that the administration route may have a significant impact, both quantitatively and qualitatively, on the immune response elicited in vaccinated animals. The delivery method may also influence the dose of DNA required to raise an effective immune response. Today, a variety of routes for gene delivery are available including intramuscular, intradermal, intravenous, intraperitoneal [20], oral [21, 22], intranasal [20, 23-25] and vaginal [26, 27]. The two most common routes for gene delivery, skin and muscle, are generally considered to have significant differences in immunocompetency [28]. Usually 10 to 100μg of plasmid DNA are required to elicit antigen specific immune responses by intramuscular or subcutaneous routes, but this amount can vary according to the antigen expressed and the animal model used. As stated before, the delivery system used can also influence the amount of DNA. Thus, for a given antigen and animal species, the best route of administration will probably have to be determined empirically.

Although partially empirical, the type of induced immune response can influence the choice of the delivery method. For example, it has been demonstrated that mucosal immunization was superior to systemic immunization to induce and sustain mucosal IgA responses [24, 25, 27]. Furthermore, mucosal vaccination can induce a stronger immune response at the site of pathogen entry, thus leading to enhanced protection [29].

It is also possible to deliver DNA vaccines using electroporation, "gene gun" and tattoo.

Electroporation (EP) is a method in which millisecond electrical pulses are applied to the vaccination site shortly after the administration of plasmid DNA. During this period, the entry of plasmid is facilitated [30, 31]. The tissue damage caused by electroporation causes inflammation and recruits antigen presenting cells to the injection site [32]. Also, EP improves cell transfection, resulting in 100-1000 fold increase in protein expression [33-35]. In summary, *in vivo* EP enhances the immunogenicity of DNA vaccines in mice [34, 36-38] and in large animal models [6, 39-41]. Furthermore, EP has been shown to increase the number of polyfunctional T cells, a feature associated with successful immune defense against infectious agents [42]. Recently, two early-phase safety and/or efficacy trials were performed to evaluate the effect of electroporation on DNA immunogenicity in humans [43] and today there are 27 ongoing clinical trials (www.clinicaltrials.gov).

Gene gun is another physical method to deliver DNA vaccines. Gene gun uses gas-driven ballistic devices that propels gold beads coated with DNA and deliver them directly into the cytosol of epidermal cells, resulting in transgene expression [44-45]. Immunization with a DNA vaccine using gene gun often requires 0.1-1 μg of plasmid DNA to induce specific immune responses. The ability of this technology to induce antibody and cell-mediated immune responses in humans has been evaluated in three clinical trials [46]. Using a DNA vaccine encoding the hepatitis B antigen these studies demonstrated that all vaccinees seroconverted, including those who had not responded to the licensed vaccine. In addition, the DNA vaccine induced significant $CD8^+$ and $CD4^+$ T cell responses in 100% of the subjects [47-48].

A recent study compared the relative immunogenicity of a DNA vaccine expressing an antigen from H5N1 influenza virus when delivered by intramuscular needle immunization, gene gun or electroporation [6]. The authors observed that both the gene gun (GG) and electroporation (EP) methods were more immunogenic than the intramuscular (IM) method. However, EP and IM stimulated a Th-1 type antibody response and the antibody response to GG was Th-2 dominated.

Another comparative study conducted used IM and GG of a DNA vaccine encoding an antigen (E7) from human papiloma virus (HPV-16) fused to *Mycobacterium tuberculosis* heat shock protein 70 (HSP70). The authors observed that GG immunization induced the highest number of antigen-specific CD8+ T cells and better anti tumor effects [49].

More recently, another study performed a head-to-head comparison of IM, EP and GG using a DNA vaccine encoding the E7 antigen from HPV-16. They demonstrated that vaccination via EP generated the highest number of E7-specific cytotoxic $CD8^+$ T cells, which correlated to improved outcomes in the treatment of growing tumors. In addition, they observed that EP resulted in significantly higher levels of circulating protein compared to GG or IM vaccination [50].

DNA tattooing is a technique that uses a perforating needle device that oscillates at a constant high frequency and punctures the skin resulting in transfection of skin-associated cells and expression of the antigen [51]. It has been demonstrated that DNA tattooing induces stronger vaccine-specific immune responses over intramuscular DNA immunization in mice [52, 53] and nonhuman primates [54].

These findings provided important information for the further selection and optimization of DNA vaccine delivery methods for human applications.

## Advantages X Limitations

As stated before, DNA vaccines are considered the safest vaccine platform available. They offer the promise of a molecularly defined reagent that is neither infectious nor capable of replication. Beyond the safety profile, DNA vaccines do not induce vector immunity and can elicit both humoral and cellular specific immune responses against the encoded antigen. DNA vaccines are easily designed, low cost manufacturing process and can be produced at commercial scale with a high degree of purity. Furthermore, DNA vaccines form stable formulations (do not require a preservative in the final preparation) that facilitate storage and shipping. Therefore, DNA plasmid technology has a number of advantages over other vaccine strategies.

Since plasmid DNA is a safe vaccine platform the main limitation for its use in humans is the low immunogenicity. It has been a challenge to transfer the success of inducing potent immunity observed in small animal models to humans. Some of the approaches now available to enhance the immunogenicity of DNA vaccines will be reviewed below. The main advantages and limitations of DNA vaccination are displayed in Table 1.

A number of safety concerns came up since the beginning of DNA vaccine utilization in the early 90′s [55]. These include the possibility that such vaccines may (i) stimulate the production of autoantibodies against the plasmid′s DNA, potentially inducing or accelerating the development of systemic autoimmune diseases; (ii) integrate into the host genome, increasing the risk of carcinogenesis or other genetic abnormalities; (iii) induce the development of tolerance rather than immunity; (iv) selectively alter host′s cytokine response to infections.

Many of such concerns have been elucidated in the past years. For example, cumulative data from clinical trials showed that DNA vaccines did not accelerate systemic or organ-specific autoimmune diseases. Based on these findings, the updated Food and Drug Administration (FDA) 2007 Guidance document concluded that sponsors no longer need to perform pre-clinical studies to specifically assess the effect of prophylactic vaccination on autoimmunity [56]. DNA vaccines intended for other uses, such as the treatment of autoimmune diseases or cancer, are not covered by the document. Furthermore, there is no evidence from pre-clinical or clinical trials that DNA vaccines result in the development of tolerance in adults.

**Table 1. Advantages x limitations of DNA vaccines**

| Advantages | Limitations |
|---|---|
| Easy design and production<br>New available molecular biology tools facilitate design and production<br>Easy to scale up<br>Large scale production methods available | Low immunogenicity<br>DNA vaccines are poorly immunogenic in humans |
| Safe vaccine platform<br>Subunit vaccine; not infectious, not replicative; unable to revert to virulent forms | |
| Stable formulation<br>Easy storage and shipping; no cold chain requirement for transport<br>Cost-effectiveness | |
| Immunogenic<br>Can induce both humoral and cellular specific immune responses (including cytotoxic T cell) | |

An overview of the specific guidelines and regulatory aspects for manufacturing, preclinical immunogenicity, safety, quality assurance and quality control of prophylactic DNA vaccines were developed by the World Health Organization (WHO) (http://www.who.int/biologicals/publications/ECBS%202005%20Annex%201%20DNA.pdf) and the FDA [56]. Recently, a paper described the evolution of FDA policy, the status of current regulatory guidance and several recommendations to facilitate the development of prophylactic DNA vaccines [55]. One main issue is that the production process should conform to cGMP (current Good Manufacturing Practices) guidelines and be acceptable to the FDA or other national regulatory agencies.

After endorsement of the quality and pre-clinical safety of a new DNA vaccine, clinical trials should proceed through three phases. Typically, phase I trials, which involve a small group of healthy volunteers (20-80), are designed to determine whether the vaccine formulation is safe and immunogenic. Phase II trials involve a larger number of healthy volunteers and are designed to further evaluate vaccine safety and potential side effects, immunogenicity, optimum dosage and schedule. Phase III trials monitor safety, potential side effects and evaluate efficacy on a large scale. These trials must be large enough (thousands of volunteers) to ensure that the vaccine works under various conditions. If Phase III results demonstrate safety and sufficient efficacy, the manufacturer applies for permission to license and market the product and submits a plan for long-term, post-licensure safety monitoring (Phase IV trials). A full set of clinical trials for a successful candidate can take 10 to 12 years, involve 50,000 to 100,000 volunteers, and cost millions of dollars. Because of these features, few vaccine candidates survive this rigorous process.

DNA vaccines have also been developed for veterinary use and efficacy in animal target species is being observed in some trials. Potentially protective immune responses are observed against many infectious agents in several target species including fish, and in companion and farm animals. A veterinary DNA vaccine to protect horses from West Nile virus was first licensed in 2005 by FDA [57]. In 2008, the Australian Pesticides and Veterinary Medicines Authority approved a DNA-based growth hormone, delivered using electroporation (EP), for use in swines [58]. In total, four animal DNA vaccines were approved for the vaccination of horses, salmon, swine and dogs (Table 2).

While the quality and safety considerations for veterinary use differ from vaccines for human use, experience with veterinary DNA vaccines can provide valuable information for human use.

**Table 2. Licensed DNA vaccines**

| Vaccine name | Organism | Vaccine target |
|---|---|---|
| West Nile Innovator | Horses | West Nile virus |
| Apex-IHN | Salmon | Infectious haematopoietic necrosis virus |
| LifeTide-SWS | Swine | Growth hormone realizing hormone (GHRH) |
| Canine Melanoma | Dogs | Melanoma |

## Strategies to Enhance Immunogenicity of DNA Vaccines

Despite a large number of clinical studies (98 studies in the world as of January 2011, see www. clinicaltrials.gov and Table 3) no DNA vaccine for humans, although safe, has yet met applicable efficacy requirements. This is due to the fact that the induced immune response in humans/non human primates by DNA vaccines is not potent as it is in small animal models. Usually high DNA doses in the milligram (mg) range are required in humans and even then the immunogenicity is reduced.

**Table 3. Clinical trials of DNA vaccines**

| Category | Number of clinical trials | Type | Disease |
|---|---|---|---|
| Allergy | 1 | Therapeutic | Seasonal allergic rhinitis |
| Cancer | 24 | Therapeutic | Breast, colorectal, kidney, prostate and liver cancer, lymphoma, melanoma, cervical, leukemia |
| Infectious diseases | 58 | Prophylactic | Herpes simplex Virus 2 (HSV-2), Dengue, Human Immunodeficiency virus (HIV), malaria, influenza, Human papiloma virus (HPV), Enterovirus 71 |
| | 15 | Therapeutic | Chronic hepatitis B, Chronic Hepatitis C, Herpes simplex Virus 2 (HSV-2), Human Immunodeficiency virus (HIV) |

**Table 4. Approaches to enhance immunogenicity of DNA vaccines**

| Approach | Example |
|---|---|
| Plasmid design | Promoter/enhancer elements |
| | Kosak sequence |
| | Leader sequences |
| | Codon optimization |
| | Cytokines |
| Delivery Methods | Electroporation |
| | Gene Gun |
| | Tatoo |
| Target to dendritic cells | DEC205 |
| Improved immunization protocols | Heterologous prime boost |
| Adjuvants | Alum |
| | Imunostimulatory complexes (ISCOMs) |
| | Bacterial compounds |
| | Cytokines |
| | Chemical compounds |

In the last years, a variety of approaches (Table 4) are under evaluation to increase the potency of DNA vaccines. Some of these are based on modifications of plasmid basic design, inclusion of adjuvants in the formulation, targeting to dendritic cells and the use of next-generation delivery methods (presented in an earlier section of this review).

## Modifications of Plasmid Basic Design

Modifications of the plasmid can greatly enhance the level of gene transcription. In general, the higher the level of target gene expression, the stronger the immune response. Optimizing the transcriptional elements in the plasmid backbone was evaluated as a strategy to improve gene transcription and expression. One important component of the plasmid is the promoter that drives expression of the gene of interest.

In general, virally-derived promoters/enhancers have provided greater gene expression in vivo than other eukaryotic promoters. In particular, the human cytomegalovirus (CMV) immediate early enhancer-promoter (known as the CMV promoter) has often been shown to direct the highest level of transgene expression in eukaryotic tissues when compared with other promoters [59]. For most vaccine plasmids, the human CMV promoter is a common choice because it promotes high-levels of constitutive expression in a wide range of mammalian cells. For example, a DNA vaccine encoding human immunodeficiency virus (HIV) genes was designed using the CMV promoter/enhancer and compared to another plasmid using the endogenous AKV murine leukemia long terminal repeat (LTR). The results showed that macaques immunized with the plasmid under the control of CMV presented higher HIV-specific immune responses [60].

A second important modification to the plasmid is the inclusion of a termination site, or poly(A) signal site, that is required for proper termination of transcription and export of the mRNA from the nucleus. The polyadenylation sequence used within a DNA vaccine may also have significant effects on transgene expression. Many DNA vaccines use the bovine growth hormone (BGH) terminator sequence or simian virus 40 (SV40) to ensure proper transcriptional termination.

Another strategy that is used to improve the expression of the gene is to optimize the initiation start site for protein synthesis, since sequences flanking the initiator codon influence its recognition by eukaryotic ribosomes. To optimize the gene, a Kosak sequence should be included on the sequence immediately upstream of the target gene´s ATG [61]. The consensus of Kosak sequence is gccRccAUGG, where R is a *purine* (*adenine* or *guanine*) three bases upstream of the *start codon* (AUG), which is followed by another 'G'. Addition of a secretory leader sequence (e.g IgE leader sequence, human tissue plasminogen activator (tPA) leader sequence) is another way to enhance antigen expression by stabilizing the mRNA and contributing to translational efficiency [62, 63].

Although sharing the same genetic code, several species have preferences for the use of particular codons. This is due to the fact that not all transfer RNAs (tRNA) exist at equal levels within cells. One of the most effective ways to increase the expression of the encoded protein is through the use of codon optimization, i.e. select codons that target more abundant tRNAs within the cell. This procedure correlates with translational efficiency in mammalian cells [64], results in increased protein production and enhanced immune responses [65-69].

In a recent work Yan et al. [70] designed a DNA vaccine encoding a HIV envelope gene with the goal of increasing vaccine antigen immune potency. The vaccine cassette was designed with several modifications, including addition of Kosak sequence, codon optimization, and a substituted immunoglobulin E leader sequence [70]. The results demonstrated that this construct was up to four times more potent at driving cellular immune responses. In addition, another work evaluated the relative contributions of codon usage, promoter efficiency and leader sequence to the antigen expression and immunogenicity of HIV-based DNA vaccine [71]. The authors observed that all these factors can work synergistically to improve the final antigen expression and immunogenicity. Moreover, the best result came from the approach that optimized all three components in a DNA vaccine design.

Another approach to power the immunogenicity of DNA immunizations is to incorporate into the plasmid vector containing the antigen sequence, genes coding for cytokines like interferon gamma (IFN-$\gamma$) [72] and tumor necrosis factor alpha (TNF$\alpha$) [73].

## Delivery Methods

The most common route for DNA immunization is the IM. However, as stated in a previous section, several new delivery methods have been developed to improve the immunogenicity of such vaccines. This new methods aim at inducing higher antigen expression (e.g. electroporation) and/or direct transfection of APCs (e.g. gene gun).

The better performance of DNA vaccines delivered with a gene gun device for inducing cellular immune responses can be attributed to the cell types involved in antigen processing and presentation. The ID route preferentially favors transfection of epidermal keratinocytes (non professional APCs) as well as APCs such as Langerhans cells [74]. For example, it has been demonstrated that injection of gold beads coated with a DNA vaccine against hepatitis B was able to induce antibodies in individuals who had not responded to the licensed recombinant vaccine [75].

## Target to Dendritic Cells

Regardless of the route of administration, the critical cell types to target with DNA vaccine may be dendritic cells (DCs), the most efficient APCs. To date various endocytic receptors have been identified whose expression is limited to DCs. Many of these are C-type lectins, (Langerin/ CD207, DC-SIGN/ CD209, BDCA-2, Dectin-1 and DCIR-2), or type I proteins with multiple lectin domains, for example, mannose receptor (MR)/CD206 and DEC-205/CD205 [76].

Recent studies have demonstrated that antigens can be targeted selectively to DCs *in vivo* when antigen is incorporated into an antibody against the DC endocytic receptor, DEC205/CD205 [77-78]. This approach was also tested using DNA vaccines [79]. Nchinda *et al.* [80] introduced the sequences for a single chain anti-DEC205 antibody into a DNA plasmid that was used to deliver a HIV antigen. They found that it greatly improved the delivery of antigen to DC *in vivo* as well as antigen presentation to $CD4^+$ and $CD8^+$ T cells

[80]. At the same time, they were able to reduce by 100-fold the amount of DNA that was required to induce T-cell immunity as well as protection.

# Improved Immunization Protocols

A very promising strategy that has been evaluated in clinical trials is to combine DNA vaccines with other vaccine modalities. In this scenario the idea is to prime the immune system to a target antigen delivered by a DNA vaccine and then boost this immunity by re-administration of the antigen in the context of a second and distinct vector (such as recombinant viruses or proteins). This strategy has become known as heterologous prime-boost and is more effective than the 'homologous' prime–boost approach [81]. The sequence of immunization is also important to induce more potent immune responses, with DNA being the optimal prime rather than any of the viral vectors [82].

The exact immune mechanisms by which heterologous prime-boost is so potent have not yet been fully elucidated [83,84]. However it has now become clear that both $CD4^+$ and $CD8^+$ T cell responses can be powered using such strategies [81]. For example, in the simian immunodeficiency virus (SIV) model, a DNA prime-adenovirus boost was able to induce powerful $CD4^+$ and $CD8^+$ T cell responses in animals that control viremia after heterologous challenge [85]. Indeed, several reports using the SIV model demonstrated the superiority of heterologous prime boost regimens in nonhuman primates [35].

Recently, a clinical trial evaluated the prime-boost approach using a DNA and *vaccinia*-based vaccines against falciparum malaria. T cell responses to the prime-boost regimen were five-to tenfold higher than those induced by either DNA vaccine or recombinant *vaccinia* virus alone [86]. In the HIV vaccine scenario, no antigen-specific antibody responses were detected in human volunteers after three DNA immunizations, but they were rapidly detected when the volunteers received a protein boost [38]. Another HIV clinical trial provided the clear evidence that a DNA prime-NYVAC poxviral vector boost was more immunogenic than the NYVAC poxviral vector alone [87]. Moreover, the prime-boost regimen elicited polyfunctional and long-lasting T cells responses, desirable features for T-cell based vaccines.

The prime–boost vaccine approach can also improve the effectiveness of existing vaccines [88]. For example, mice primed with DNA and boosted with the licensed hepatitis B protein vaccine presented stronger and more homogenous antibody responses when compared to groups immunized with the recombinant protein only [89].

# Adjuvants

Adjuvants (from the latin adjuvare, to help) are substances that can be co-administered with a vaccine to accelerate, prolong or enhance the quality of an immune response to the antigen. These substances, that encompass a wide range of compounds, are subject of clinical and experimental studies in order to induce effective and protective immune response, mainly immunological memory to diseases whose traditional vaccines approaches has proven to be insufficient (AIDS, malaria, trypanosomiasis, and cancer). In the last years several studies aimed to understand the molecular mechanisms by which adjuvants mediated their functions on innate and adaptative cells of the immune system, improving vaccine efficacy, especially those that the effective response is based on CTL or TCD4+ Th1 responses. Among these vaccines are included the DNA vaccines, classified as 3rd vaccine generation, the focus of present chapter. As stated before, these vaccines induce both humoral and cellular immunity, however as demonstrated in experimental and clinical studies [16], high doses (mg range) of DNA are necessary to detect immune response in humans, showing its low immunogenicity. Due this fact, it is clear or even mandatory the use of adjuvants associated with this type of vaccine. Currently, adjuvant selection brings a major breakthrough for the use of DNA vaccines for both prophylaxis and therapeutic treatment.

Different categories of adjuvants (Table 4), based on their nature have been studied, but few has advanced for use in humans, in part by the difficulty to license them according to the Guideline on Adjuvants in Vaccines for Human use by The European Medicines Agency or by FDA.

One of the first adjuvant described and one of the few licensed worldwide for human use is Alum that is composed by aluminum salts (aluminum phosphate and aluminum hydroxide).The mechanism by which Alum exerts their adjuvant effect is until now under investigation but it is well known that Alum precipitates on tissues when inoculated with the antigen allowing its slow release and recognition by immunocompetent cells. It was recently demonstrated *in vitro* that Alum can activate NLRP3/inflamosome that induces mature IL-1 beta [90]. Inflammasome activation could explain the inflammation induced by Alum and the mixed Th2/Th1 pattern observed in humans, but does not explain the strong Th2 response in mice when this substance is used [91,92]. Several studies using DNA vaccine associated with Alum has demonstrated the enhancement in antibodies levels [93-95].

Other adjuvant formulation that is already licensed for human use in Europe is MF59 an oil-in-water emulsion based on squalene. This oil (squalene) is metabolized more rapidly than those derived from paraffin found in Freund's adjuvant [91]. The mechanism by which this mixture exerts their adjuvant effect is unclear. It is known that MF59 increases antibodies levels; induces a mixed Th1/Th2 pattern of CD4 T cell responses [96]; increases APC and granulocyte migration to the site of antigen inoculation and promotes differentiation of monocytes into mature dendritic cells [97]. MF59 is also employed to increase the potency of DNA vaccines [98].

Another category of adjuvant which allows better capture of antigen by APCs is the Imunostimulatory Complexes (ISCOMs). ISCOMs are composed of saponins (derived from the bark tree *Quillaja saponaria*) formulated with cholesterol, phospholipid and antigen. Like other adjuvants, ISCOMs are able to induce antibody, T helper and CTL responses [99] in several species, including non-human primates [100] and humans [101]. In addition it seems that ISCOMs act at the endosomal membrane allowing escape of antigen into the cytoplasm and hence their presentation by MHC class I [102]. Coadministration of DNA vaccines with ISCOMs had already been described [103,104].

The knowledge of the receptors (pathogen recognition receptors, PRRs) involved in innate immunity that directly influences the adaptive immune response, has implemented the study of adjuvants. PRRs consist of nonphagocytic receptors such as Toll-like receptors (TLRs). These receptors are found on the cell surface and in endosomes and when activated lead to a signal transduction cascade. TLRs can recognize pathogen associated molecular patterns (PAMPs) and its natural ligands or synthetic agonists have been used as adjuvants [105]. Also, TLRs agonists have been used to improve DNA vaccines efficacy [94,106,107].

Intrinsic adjuvant effect mediated by TLRs can be observed in most DNA vaccines due the presence of unmethylated cytosine-phosphate-guanine oligonucleotide sequences (CpG motifs) that binds to TLR-9, as stated in an earlier section. Synthetic oligodeoxynucleotides (ODN) containing such CPG motifs are also been studied as a soluble adjuvant to improve experimental DNA immunization against leishmaniasis [108] and SIV [94].

It has long been described that killed-bacteria suspensions such as Bacille Calmette Guerin (BCG), *Propionibacterium acnes* and *Bordetella pertussis* have adjuvant effect. Undoubtedly much of their effects are mediated by TLR activation. The purified and detoxified bacteria component, monophosphoril lipid A (MPL), is already in use in human vaccines. This component binds to TLR-4 and activates the TRIF signaling pathway, leading to cell activation. Both licensed HBV and papilloma (Cervarix-GSK) vaccines contains MPL formulated with antigens and Alum (AS04) [109].

Another bacterium with adjuvant property is *Propionibacterium acnes* (in the past classified as *Corynebacterium parvum*). We used killed-*Propionibacterium acnes* suspension or its soluble polysaccharide compound as adjuvant [110], in several murine experimental models. We observed increase in Th1 immune response including high levels of specific IgG2a and IFN gamma synthesis when mice were immunized with a DNA vaccine encoding a *Trypanosoma cruzi* antigen. This adjuvanted DNA vaccine was also able to reduce peak parasitemia after *T. cruzi* challenge [111]. One of the possible mechanisms by which both whole bacterial suspension and purified polysaccharide from *P.acnes* increase the immune response is dendritic cell (DC) activation. We demonstrated that both *P. acnes* preparations were able to increase *in vivo* and *in vitro* DC maturation [110]. Also, it has been demonstrated that *P.acnes* exert its immunomodulatory effects by binding to TLR-2 and TLR-9 [112-114].

Another approach to improve the immunogenicity of DNA immunization is to incorporate into the plasmid vector genes coding for cytokines, chemokines, costimulatory molecules and antiapoptotic genes as has been reviewed elsewhere [115-117]. It has been demonstrated that addition of such adjuvants may increase the breadth and magnitude of immune response and skew the type of immune response [10]. Of note, cytokines such as interleukin (IL)-12 and IL-15 have been effective in enhancing the immune response in both murine and nonhuman primate models [118,119]. This approach has been used successfully against tuberculosis [120,121], HIV [122] and cytomegalovirus [123].

Chemical compounds have also been evaluated as adjuvants for DNA vaccines [124]. One of such compounds is Bupivacaine, a local anesthetic drug belonging to the *amino amide* group that blocks neuron transmission. Bupivacaine is also a myotoxin that when injected destroys myofiber cells leading to the clearence of cell debris and proliferation of myoblasts. Pretreatment of muscle with bupivacaine several days prior to injection of DNA, results in increased DNA uptake, as evidenced by increased DNA expression at the injection site [26,125]. Furthermore, intramuscular immunization with bupivacaine:DNA complexes results in greater immune responses [126]. Recently, Jamali et al. [127] evaluated the use of naloxone, a general opioid antagonist, as an adjuvant for a DNA vaccine encoding the glycoprotein-1 from herpes simplex virus-1 (HSV-1). It was observed that administration of naloxone increased cellular immune responses, shifted the immune response toward a Th1 pattern and improved protection against HSV-1.

## Conclusion

DNA vaccination holds great promise in both prophylactic and therapeutic vaccines. Since the early 1990s DNA vaccines have been evaluated as a potential rational vaccine platform that could provide several advantages over conventional vaccine approaches. In particular, DNA immunization could induce protective humoral and cellular immune responses (including cytotoxic T cells) against infectious diseases, allergy and tumors. Furthermore, DNA vaccines offer a number of significant advantages over traditional vaccine modalities such as safety, stability, easy design and feasible production method. Even though DNA vaccines present several advantages, their efficacy in humans has proven marginal. A wide variety of strategies have been evaluated to increase the potency of such vaccines including modifications of plasmid design, new delivery methods, addition of adjuvants and targeting to dendritic cells. Furthermore, mixed immunization protocols are promising. Due to recent advances in the understanding of DNA vaccines immunology, we believe that in the near future this vaccine modality will become effective in human clinical settings.

## References

[1] Kutzler MA, Weiner DB (2008) DNA vaccines: ready for prime time? Nat. Rev. Genet. 9: 776-788.

[2] Ledgerwood JE, Graham BS (2009) DNA vaccines: a safe and efficient platform technology for responding to emerging infectious diseases. Hum. Vaccin. 5: 623-626.

[3] Lu S (2008) Immunogenicity of DNA vaccines in humans: it takes two to tango. Hum. Vaccin 4: 449-452.

[4] Rice J, Ottensmeier CH, Stevenson FK (2008) DNA vaccines: precision tools for activating effective immunity against cancer. Nat. Rev. Cancer 8: 108-120.

[5] Liu MA (2010) DNA vaccines: an historical perspective and view to the future. Immunol. Rev. 239: 62-84.

[6] Wang S, Zhang C, Zhang L, Li J, Huang Z, et al. (2008) The relative immunogenicity of DNA vaccines delivered by the intramuscular needle injection, electroporation and gene gun methods. Vaccine 26: 2100-2110.

[7] Germain RN (2010) Vaccines and the future of human immunology. Immunity 33: 441-450.

[8] Tang DC, DeVit M, Johnston SA (1992) Genetic immunization is a simple method for eliciting an immune response. Nature 356: 152-154.

[9] Apostolopoulos V, Weiner DB (2009) Development of more efficient and effective DNA vaccines. Expert Rev. Vaccines 8: 1133-1134.

[10] Abdulhaqq SA, Weiner DB (2008) DNA vaccines: developing new strategies to enhance immune responses. Immunol. Res. 42: 219-232.

[11] Schirmbeck R, Reimann J (2001) Revealing the potential of DNA-based vaccination: lessons learned from the hepatitis B virus surface antigen. Biol. Chem. 382: 543-552.

[12] Hemmi H, Takeuchi O, Kawai T, Kaisho T, Sato S, et al. (2000) A Toll-like receptor recognizes bacterial DNA. Nature 408: 740-745.

[13] Leifert JA, Whitton, J.L. (2000) Immune Responses to DNA Vaccines: Induction of CD8 T Cells. Madame Curie Bioscience Database, Landes Bioscience
[14] Condon C, Watkins SC, Celluzzi CM, Thompson K, Falo LD, Jr. (1996) DNA-based immunization by in vivo transfection of dendritic cells. Nat. Med. 2: 1122-1128.
[15] Bot A, Stan AC, Inaba K, Steinman R, Bona C (2000) Dendritic cells at a DNA vaccination site express the encoded influenza nucleoprotein and prime MHC class I-restricted cytolytic lymphocytes upon adoptive transfer. Int. Immunol. 12: 825-832.
[16] Liu MA (2003) DNA vaccines: a review. J. Intern. Med. 253: 402-410.
[17] Albert ML, Sauter B, Bhardwaj N (1998) Dendritic cells acquire antigen from apoptotic cells and induce class I-restricted CTLs. Nature 392: 86-89.
[18] Feltquate DM, Heaney S, Webster RG, Robinson HL (1997) Different T helper cell types and antibody isotypes generated by saline and gene gun DNA immunization. J. Immunol. 158: 2278-2284.
[19] Torres CA, Iwasaki A, Barber BH, Robinson HL (1997) Differential dependence on target site tissue for gene gun and intramuscular DNA immunizations. J. Immunol. 158: 4529-4532.
[20] Fynan EF, Webster RG, Fuller DH, Haynes JR, Santoro JC, et al. (1993) DNA vaccines: protective immunizations by parenteral, mucosal, and gene-gun inoculations. Proc. Natl. Acad. Sci. U S A 90: 11478-11482.
[21] Roy K, Mao HQ, Huang SK, Leong KW (1999) Oral gene delivery with chitosan--DNA nanoparticles generates immunologic protection in a murine model of peanut allergy. Nat. Med. 5: 387-391.
[22] Herrmann JE, Chen SC, Jones DH, Tinsley-Bown A, Fynan EF, et al. (1999) Immune responses and protection obtained by oral immunization with rotavirus VP4 and VP7 DNA vaccines encapsulated in microparticles. Virology 259: 148-153.
[23] Klavinskis LS, Gao L, Barnfield C, Lehner T, Parker S (1997) Mucosal immunization with DNA-liposome complexes. Vaccine 15: 818-820.
[24] Sasaki S, Hamajima K, Fukushima J, Ihata A, Ishii N, et al. (1998) Comparison of intranasal and intramuscular immunization against human immunodeficiency virus type 1 with a DNA-monophosphoryl lipid A adjuvant vaccine. Infect. Immun. 66: 823-826.
[25] Sasaki S, Sumino K, Hamajima K, Fukushima J, Ishii N, et al. (1998) Induction of systemic and mucosal immune responses to human immunodeficiency virus type 1 by a DNA vaccine formulated with QS-21 saponin adjuvant via intramuscular and intranasal routes. J. Virol. 72: 4931-4939.
[26] Wang B, Dang K, Agadjanyan MG, Srikantan V, Li F, et al. (1997) Mucosal immunization with a DNA vaccine induces immune responses against HIV-1 at a mucosal site. Vaccine 15: 821-825.
[27] Livingston JB, Lu S, Robinson H, Anderson DJ (1998) Immunization of the female genital tract with a DNA-based vaccine. Infect. Immun. 66: 322-329.
[28] Thomas Tüting JA, Walter J. Storkus and Louis D. Falo (2000) The Immunology of DNA Vaccines; Lowrie DB, Whalen, R.G., editor. Totowa, NJ: Humana Press Inc.
[29] Amorij JP, Hinrichs W, Frijlink HW, Wilschut JC, Huckriede (2010) A Needle-free influenza vaccination. Lancet Infect. Dis. 10: 699-711.
[30] Tieleman DP (2004) The molecular basis of electroporation. BMC Biochem 5: 10.
[31] Tarek M (2005) Membrane electroporation: a molecular dynamics simulation. Biophys. J. 88: 4045-4053.

[32] Liu J, Kjeken R, Mathiesen I, Barouch DH (2008) Recruitment of antigen-presenting cells to the site of inoculation and augmentation of human immunodeficiency virus type 1 DNA vaccine immunogenicity by in vivo electroporation. J. Virol. 82: 5643-5649.

[33] Rizzuto G, Cappelletti M, Maione D, Savino R, Lazzaro D, et al. (1999) Efficient and regulated erythropoietin production by naked DNA injection and muscle electroporation. Proc. Natl. Acad. Sci. U S A 96: 6417-6422.

[34] Widera G, Austin M, Rabussay D, Goldbeck C, Barnett SW, et al. (2000) Increased DNA vaccine delivery and immunogenicity by electroporation in vivo. J. Immunol. 164: 4635-4640.

[35] Khan AS, Pope MA, Draghia-Akli R (2005) Highly efficient constant-current electroporation increases in vivo plasmid expression. DNA Cell Biol. 24: 810-818.

[36] Dobano C, Widera G, Rabussay D, Doolan DL (2007) Enhancement of antibody and cellular immune responses to malaria DNA vaccines by in vivo electroporation. Vaccine 25: 6635-6645.

[37] Hooper JW, Golden JW, Ferro AM, King AD (2007) Smallpox DNA vaccine delivered by novel skin electroporation device protects mice against intranasal poxvirus challenge. Vaccine 25: 1814-1823.

[38] Wang S, Kennedy JS, West K, Montefiori DC, Coley S, et al. (2008) Cross-subtype antibody and cellular immune responses induced by a polyvalent DNA prime-protein boost HIV-1 vaccine in healthy human volunteers. Vaccine 26: 3947-3957.

[39] Laddy DJ, Yan J, Khan AS, Andersen H, Cohn A, et al. (2009) Electroporation of synthetic DNA antigens offers protection in nonhuman primates challenged with highly pathogenic avian influenza virus. J. Virol. 83: 4624-4630.

[40] Laddy DJ, Yan J, Kutzler M, Kobasa D, Kobinger GP, et al. (2008) Heterosubtypic protection against pathogenic human and avian influenza viruses via in vivo electroporation of synthetic consensus DNA antigens. PLoS One 3: e2517.

[41] Luckay A, Sidhu MK, Kjeken R, Megati S, Chong SY, et al. (2007) Effect of plasmid DNA vaccine design and in vivo electroporation on the resulting vaccine-specific immune responses in rhesus macaques. J. Virol. 81: 5257-5269.

[42] Brave A, Nystrom S, Roos AK, Applequist SE (2010) Plasmid DNA vaccination using skin electroporation promotes poly-functional CD4 T-cell responses. Immunol. Cell Biol.

[43] Tjelle TE, Rabussay, D., Ottensmeier, C., Mathiesen, I., Kjeken, R. (2008) Taking Electroporation-Based Delivery of DNA Vaccination into Humans: A Generic Clinical Protocol In: Li S, editor. Electroporation Protocols-Preclinical and Clinical Gene Medicine. pp. 497-507.

[44] Yang NS, Burkholder J, Roberts B, Martinell B, McCabe D (1990) In vivo and in vitro gene transfer to mammalian somatic cells by particle bombardment. Proc. Natl. Acad. Sci. U S A 87: 9568-9572.

[45] Williams RS, Johnston SA, Riedy M, DeVit MJ, McElligott SG, et al. (1991) Introduction of foreign genes into tissues of living mice by DNA-coated microprojectiles. Proc. Natl. Acad. Sci. U S A 88: 2726-2730.

[46] Fuller DH, Loudon P, Schmaljohn C (2006) Preclinical and clinical progress of particle-mediated DNA vaccines for infectious diseases. Methods 40: 86-97.

[47] Roy MJ, Wu MS, Barr LJ, Fuller JT, Tussey LG, et al. (2000) Induction of antigen-specific CD8+ T cells, T helper cells, and protective levels of antibody in humans by

particle-mediated administration of a hepatitis B virus DNA vaccine. Vaccine 19: 764-778.

[48] Swain WE, Heydenburg Fuller D, Wu MS, Barr LJ, Fuller JT, et al. (2000) Tolerability and immune responses in humans to a PowderJect DNA vaccine for hepatitis B. Dev. Biol. (Basel) 104: 115-119.

[49] Trimble C, Lin CT, Hung CF, Pai S, Juang J, et al. (2003) Comparison of the CD8+ T cell responses and antitumor effects generated by DNA vaccine administered through gene gun, biojector, and syringe. Vaccine 21: 4036-4042.

[50] Best SR, Peng S, Juang CM, Hung CF, Hannaman D, et al. (2009) Administration of HPV DNA vaccine via electroporation elicits the strongest CD8+ T cell immune responses compared to intramuscular injection and intradermal gene gun delivery. Vaccine 27: 5450-5459.

[51] van den Berg JH, Nujien B, Beijnen JH, Vincent A, van Tinteren H, et al. (2009) Optimization of intradermal vaccination by DNA tattooing in human skin. Hum. Gene Ther. 20: 181-189.

[52] Bins AD, Jorritsma A, Wolkers MC, Hung CF, Wu TC, et al. (2005) A rapid and potent DNA vaccination strategy defined by in vivo monitoring of antigen expression. Nat. Med. 11: 899-904.

[53] Pokorna D, Rubio I, Muller M (2008) DNA-vaccination via tattooing induces stronger humoral and cellular immune responses than intramuscular delivery supported by molecular adjuvants. Genet. Vaccines Ther. 6: 4.

[54] Verstrepen BE, Bins AD, Rollier CS, Mooij P, Koopman G, et al. (2008) Improved HIV-1 specific T-cell responses by short-interval DNA tattooing as compared to intramuscular immunization in non-human primates. Vaccine 26: 3346-3351.

[55] Klinman DM, Klaschik S, Tross D, Shirota H, Steinhagen F (2010) FDA guidance on prophylactic DNA vaccines: analysis and recommendations. Vaccine 28: 2801-2805.

[56] (2005) CBER/FDA.Draft Guidance for Industry: Considerations for Plasmid DNA Vaccine for Infectious Disease Indications. Guidance document.

[57] (2005) Fort Dodge Animal Health Announces Approval of West Nile Virus DNA Vaccine for Horses. PR Newswire.

[58] Person R, Bodles-Brakhop AM, Pope MA, Brown PA, Khan AS, et al. (2008) Growth hormone-releasing hormone plasmid treatment by electroporation decreases offspring mortality over three pregnancies. Mol. Ther. 16: 1891-1897.

[59] Garmory HS, Brown KA, Titball RW (2003) DNA vaccines: improving expression of antigens. Genet. Vaccines Ther 1: 2.

[60] Galvin TA, Muller J, Khan AS (2000) Effect of different promoters on immune responses elicited by HIV-1 gag/env multigenic DNA vaccine in Macaca mulatta and Macaca nemestrina. Vaccine 18: 2566-2583.

[61] Kozak M (1990) Downstream secondary structure facilitates recognition of initiator codons by eukaryotic ribosomes. Proc. Natl. Acad. Sci. U S A 87: 8301-8305.

[62] Yang JS, Kim JJ, Hwang D, Choo AY, Dang K, et al. (2001) Induction of potent Th1-type immune responses from a novel DNA vaccine for West Nile virus New York isolate (WNV-NY1999). J. Infect. Dis. 184: 809-816.

[63] Laddy DJ, Yan J, Corbitt N, Kobasa D, Kobinger GP, et al. (2007) Immunogenicity of novel consensus-based DNA vaccines against avian influenza. Vaccine 25: 2984-2989.

[64] Gustafsson C, Govindarajan S, Minshull J (2004) Codon bias and heterologous protein expression. Trends Biotechnol. 22: 346-353.

[65] Deml L, Bojak A, Steck S, Graf M, Wild J, et al. (2001) Multiple effects of codon usage optimization on expression and immunogenicity of DNA candidate vaccines encoding the human immunodeficiency virus type 1 Gag protein. J. Virol. 75: 10991-11001.

[66] Narum DL, Kumar S, Rogers WO, Fuhrmann SR, Liang H, et al. (2001) Codon optimization of gene fragments encoding Plasmodium falciparum merzoite proteins enhances DNA vaccine protein expression and immunogenicity in mice. Infect. Immun. 69: 7250-7253.

[67] Ramakrishna L, Anand KK, Mohankumar KM, Ranga U (2004) Codon optimization of the tat antigen of human immunodeficiency virus type 1 generates strong immune responses in mice following genetic immunization. J. Virol. 78: 9174-9189.

[68] Frelin L, Ahlen G, Alheim M, Weiland O, Barnfield C, et al. (2004) Codon optimization and mRNA amplification effectively enhances the immunogenicity of the hepatitis C virus nonstructural 3/4A gene. Gene Ther. 11: 522-533.

[69] Gao F, Li Y, Decker JM, Peyerl FW, Bibollet-Ruche F, et al. (2003) Codon usage optimization of HIV type 1 subtype C gag, pol, env, and nef genes: in vitro expression and immune responses in DNA-vaccinated mice. AIDS Res. Hum. Retroviruses 19: 817-823.

[70] Yan J, Yoon H, Kumar S, Ramanathan MP, Corbitt N, et al. (2007) Enhanced cellular immune responses elicited by an engineered HIV-1 subtype B consensus-based envelope DNA vaccine. Mol. Ther. 15: 411-421.

[71] Wang S, Farfan-Arribas DJ, Shen S, Chou TH, Hirsch A, et al. (2006) Relative contributions of codon usage, promoter efficiency and leader sequence to the antigen expression and immunogenicity of HIV-1 Env DNA vaccine. Vaccine 24: 4531-4540.

[72] Nimal S, McCormick AL, Thomas MS, Heath AW (2005) An interferon gamma-gp120 fusion delivered as a DNA vaccine induces enhanced priming. Vaccine 23: 3984-3990.

[73] Nimal S, Heath AW, Thomas MS (2006) Enhancement of immune responses to an HIV gp120 DNA vaccine by fusion to TNF alpha cDNA. Vaccine 24: 3298-3308.

[74] Porgador A, Irvine KR, Iwasaki A, Barber BH, Restifo NP, et al. (1998) Predominant role for directly transfected dendritic cells in antigen presentation to CD8+ T cells after gene gun immunization. J. Exp. Med. 188: 1075-1082.

[75] Rottinghaus ST, Poland GA, Jacobson RM, Barr LJ, Roy MJ (2003) Hepatitis B DNA vaccine induces protective antibody responses in human non-responders to conventional vaccination. Vaccine 21: 4604-4608.

[76] Boscardin SB, Nussenzweig M.C. ; Trumpfheller, C. ; Steiman, RM (2009) Vaccines based on dendritic cell biology.; Levine MM, editor. New York: Informa Healthcare.

[77] Hawiger D, Inaba K, Dorsett Y, Guo M, Mahnke K, et al. (2001) Dendritic cells induce peripheral T cell unresponsiveness under steady state conditions in vivo. J. Exp. Med. 194: 769-779.

[78] Boscardin SB, Hafalla JC, Masilamani RF, Kamphorst AO, Zebroski HA, et al. (2006) Antigen targeting to dendritic cells elicits long-lived T cell help for antibody responses. J. Exp. Med. 203: 599-606.

[79] Demangel C, Zhou J, Choo AB, Shoebridge G, Halliday GM, et al. (2005) Single chain antibody fragments for the selective targeting of antigens to dendritic cells. Mol. Immunol. 42: 979-985.

[80] Nchinda G, Kuroiwa J, Oks M, Trumpfheller C, Park CG, et al. (2008) The efficacy of DNA vaccination is enhanced in mice by targeting the encoded protein to dendritic cells. J. Clin. Invest. 118: 1427-1436.

[81] Woodland DL (2004) Jump-starting the immune system: prime-boosting comes of age. Trends Immunol. 25: 98-104.

[82] Schneider J, Gilbert SC, Blanchard TJ, Hanke T, Robson KJ, et al. (1998) Enhanced immunogenicity for CD8+ T cell induction and complete protective efficacy of malaria DNA vaccination by boosting with modified vaccinia virus Ankara. Nat. Med. 4: 397-402.

[83] Liu MA (2010) Immunologic basis of vaccine vectors. Immunity 33: 504-515.

[84] Liu MA (2010) Gene-based vaccines: Recent developments. Curr. Opin. Mol. Ther. 12: 86-93.

[85] Wilson NA, Keele BF, Reed JS, Piaskowski SM, MacNair CE, et al. (2009) Vaccine-induced cellular responses control simian immunodeficiency virus replication after heterologous challenge. J. Virol. 83: 6508-6521.

[86] McConkey SJ, Reece WH, Moorthy VS, Webster D, Dunachie S, et al. (2003) Enhanced T-cell immunogenicity of plasmid DNA vaccines boosted by recombinant modified vaccinia virus Ankara in humans. Nat. Med. 9: 729-735.

[87] Harari A, Bart PA, Stohr W, Tapia G, Garcia M, et al. (2008) An HIV-1 clade C DNA prime, NYVAC boost vaccine regimen induces reliable, polyfunctional, and long-lasting T cell responses. J. Exp. Med. 205: 63-77.

[88] Lu S (2009) Heterologous prime-boost vaccination. Curr. Opin. Immunol. 21: 346-351.

[89] Xiao-wen H, Shu-han S, Zhen-lin H, Jun L, Lei J, et al. (2005) Augmented humoral and cellular immune responses of a hepatitis B DNA vaccine encoding HBsAg by protein boosting. Vaccine 23: 1649-1656.

[90] Li H, Nookala S, Re F (2007) Aluminum hydroxide adjuvants activate caspase-1 and induce IL-1beta and IL-18 release. J. Immunol. 178: 5271-5276.

[91] Coffman RL, Sher A, Seder RA (2010) Vaccine adjuvants: putting innate immunity to work. Immunity 33: 492-503.

[92] Franchi L, Nunez G (2008) The Nlrp3 inflammasome is critical for aluminium hydroxide-mediated IL-1beta secretion but dispensable for adjuvant activity. Eur. J. Immunol. 38: 2085-2089.

[93] Ulmer JB, DeWitt CM, Chastain M, Friedman A, Donnelly JJ, et al. (1999) Enhancement of DNA vaccine potency using conventional aluminum adjuvants. Vaccine 18: 18-28.

[94] Kwissa M, Amara RR, Robinson HL, Moss B, Alkan S, et al. (2007) Adjuvanting a DNA vaccine with a TLR9 ligand plus Flt3 ligand results in enhanced cellular immunity against the simian immunodeficiency virus. J. Exp. Med. 204: 2733-2746.

[95] Liang ZW, Ren H, Lang YH, Li YG (2004) Enhancement of a hepatitis B DNA vaccine potency using aluminum phosphate in mice. Zhonghua Gan Zang Bing Za Zhi 12: 79-81.

[96] Ott G, Barchfeld GL, Chernoff D, Radhakrishnan R, van Hoogevest P, et al. (1995) MF59. Design and evaluation of a safe and potent adjuvant for human vaccines. Pharm. Biotechnol. 6: 277-296.
[97] Seubert A, Monaci E, Pizza M, O'Hagan DT, Wack A (2008) The adjuvants aluminum hydroxide and MF59 induce monocyte and granulocyte chemoattractants and enhance monocyte differentiation toward dendritic cells. J. Immunol. 180: 5402-5412.
[98] Ott G, Singh M, Kazzaz J, Briones M, Soenawan E, et al. (2002) A cationic sub-micron emulsion (MF59/DOTAP) is an effective delivery system for DNA vaccines. J. Control. Release 79: 1-5.
[99] Maraskovsky E, Schnurr M, Wilson NS, Robson NC, Boyle J, et al. (2009) Development of prophylactic and therapeutic vaccines using the ISCOMATRIX adjuvant. Immunol. Cell Biol. 87: 371-376.
[100] Sjolander A, Cox JC, Barr IG (1998) ISCOMs: an adjuvant with multiple functions. J. Leukoc. Biol. 64: 713-723.
[101] Rimmelzwaan GF, Baars M, van Amerongen G, van Beek R, Osterhaus AD (2001) A single dose of an ISCOM influenza vaccine induces long-lasting protective immunity against homologous challenge infection but fails to protect Cynomolgus macaques against distant drift variants of influenza A (H3N2) viruses. Vaccine 20: 158-163.
[102] Schnurr M, Orban M, Robson NC, Shin A, Braley H, et al. (2009) ISCOMATRIX adjuvant induces efficient cross-presentation of tumor antigen by dendritic cells via rapid cytosolic antigen delivery and processing via tripeptidyl peptidase II. J. Immunol. 182: 1253-1259.
[103] Le TT, Drane D, Malliaros J, Cox JC, Rothel L, et al. (2001) Cytotoxic T cell polyepitope vaccines delivered by ISCOMs. Vaccine 19: 4669-4675.
[104] Singh M, Srivastava I (2003) Advances in vaccine adjuvants for infectious diseases. Curr. HIV Res. 1: 309-320.
[105] Pashine A, Valiante NM, Ulmer JB (2005) Targeting the innate immune response with improved vaccine adjuvants. Nat. Med. 11: S63-68.
[106] Otero M, Calarota SA, Felber B, Laddy D, Pavlakis G, et al. (2004) Resiquimod is a modest adjuvant for HIV-1 gag-based genetic immunization in a mouse model. Vaccine 22: 1782-1790.
[107] Wille-Reece U, Flynn BJ, Lore K, Koup RA, Miles AP, et al. (2006) Toll-like receptor agonists influence the magnitude and quality of memory T cell responses after prime-boost immunization in nonhuman primates. J. Exp. Med. 203: 1249-1258.
[108] Ferreira JH, Gentil LG, Dias SS, Fedeli CE, Katz S, et al. (2008) Immunization with the cysteine proteinase Ldccys1 gene from Leishmania (Leishmania) chagasi and the recombinant Ldccys1 protein elicits protective immune responses in a murine model of visceral leishmaniasis. Vaccine 26: 677-685.
[109] Casella CR, Mitchell TC (2008) Putting endotoxin to work for us: monophosphoryl lipid A as a safe and effective vaccine adjuvant. Cell Mol. Life Sci. 65: 3231-3240.
[110] Squaiella CC, Ananias RZ, Mussalem JS, Braga EG, Rodrigues EG, et al. (2006) In vivo and in vitro effect of killed Propionibacterium acnes and its purified soluble polysaccharide on mouse bone marrow stem cells and dendritic cell differentiation. Immunobiology 211: 105-116.
[111] Mussalem JS, Vasconcelos JR, Squaiella CC, Ananias RZ, Braga EG, et al. (2006) Adjuvant effect of the Propionibacterium acnes and its purified soluble polysaccharide

on the immunization with plasmidial DNA containing a Trypanosoma cruzi gene. Microbiol. Immunol. 50: 253-263.

[112] Kalis C, Gumenscheimer M, Freudenberg N, Tchaptchet S, Fejer G, et al. (2005) Requirement for TLR9 in the immunomodulatory activity of Propionibacterium acnes. J. Immunol. 174: 4295-4300.

[113] Romics L, Jr., Dolganiuc A, Velayudham A, Kodys K, Mandrekar P, et al. (2005) Toll-like receptor 2 mediates inflammatory cytokine induction but not sensitization for liver injury by Propioni-bacterium acnes. J. Leukoc. Biol. 78: 1255-1264.

[114] Mussalem JS, Yendo, T.M., Squaiella, C.C., Longo-Maugéri, I.M. Killed Propionibacterium acnes modulation on TLRs,co-stimulatory, MHC II molecules expression and cytokines synthesis of B1 lymphocytes subsets from mice peritoneal exudates cells; 2008; Lisboa-Portugal. pp. 200.

[115] Barouch DH, Letvin NL, Seder RA (2004) The role of cytokine DNAs as vaccine adjuvants for optimizing cellular immune responses. Immunol. Rev. 202: 266-274.

[116] Stevenson FK (2004) DNA vaccines and adjuvants. Immunol. Rev. 199: 5-8.

[117] Lu S (2009) Gene-based adjuvants: a new meaning. Hum. Gene Ther. 20: 1101-1102.

[118] Morrow MP, Weiner DB (2008) Cytokines as adjuvants for improving anti-HIV responses. AIDS 22: 333-338.

[119] Kraynyak KA, Kutzler MA, Cisper NJ, Laddy DJ, Morrow MP, et al. (2009) Plasmid-encoded interleukin-15 receptor alpha enhances specific immune responses induced by a DNA vaccine in vivo. Hum. Gene Ther. 20: 1143-1156.

[120] Okada M, Kita Y, Nakajima T, Kanamaru N, Hashimoto S, et al. (2007) Evaluation of a novel vaccine (HVJ-liposome/HSP65 DNA+IL-12 DNA) against tuberculosis using the cynomolgus monkey model of TB. Vaccine 25: 2990-2993.

[121] Okada M, Kita Y (2008)Tuberculosis vaccine development: The development of novel (preclinical) DNA vaccine. Hum. Vaccin 6: 297-308.

[122] Hirao LA, Wu L, Khan AS, Hokey DA, Yan J, et al. (2008) Combined effects of IL-12 and electroporation enhances the potency of DNA vaccination in macaques. Vaccine 26: 3112-3120.

[123] Cull VS, Broomfield S, Bartlett EJ, Brekalo NL, James CM (2002) Coimmunisation with type I IFN genes enhances protective immunity against cytomegalovirus and myocarditis in gB DNA-vaccinated mice. Gene Ther.9: 1369-1378.

[124] Jin H, Li Y, Ma Z, Zhang F, Xie Q, et al. (2004) Effect of chemical adjuvants on DNA vaccination. Vaccine 22: 2925-2935.

[125] Vitadello M, Schiaffino MV, Picard A, Scarpa M, Schiaffino S (1994) Gene transfer in regenerating muscle. Hum. Gene Ther 5: 11-18.

[126] Pachuk CJ, Ciccarelli RB, Samuel M, Bayer ME, Troutman RD, et al. (2000) Characterization of a new class of DNA delivery complexes formed by the local anesthetic bupivacaine. Biochim. Biophys. Acta 1468: 20-30.

[127] Jamali A, Mahdavi M, Hassan ZM, Sabahi F, Farsani MJ, et al. (2009) A novel adjuvant, the general opioid antagonist naloxone, elicits a robust cellular immune response for a DNA vaccine. Int. Immunol. 21: 217-225.

In: DNA Vaccines: Types, Advantages and Limitations
Editors: E. C. Donnelly and A. M. Dixon
ISBN 978-1-61324-444-9


Chapter V

# Co-Administration of Gamma Interferon Gene with DNA Vaccine Expressing Duck Hepatitis B Virus (DHBV) Proteins Enhances Therapeutic Efficacy of DNA-Based Immunization in Chronic Virus Carriers

***Fadi Saade,[1,2] Thierry Buronfosse,[1,2,3] Ghada Khawaja,[1,2,4] Sylviane Guerret,[5] Catherine Jamard,[1,2] Michèle Chevallier,[6] Pierre Pradat,[7] Fabien Zoulim[1,2,7] and Lucyna Cova[1,2,*]***

[1]Université Claude Bernard Lyon 1, F-69008, France
[2]Equipe 15, Centre de Recherche en Cancérologie de Lyon, UMR Inserm 1052, Lyon F69424, France,
[3]VetAgro-Sup, Marcy l'Etoile F-69280, France
[4]ISPB -Faculté de Pharmacie, Lyon F-69008, France
[5]NOVOTEC, Lyon F69007, France;
[6]Laboratoires Marcel Mérieux, Lyon 69007, France;
[7]Department of Hepatology, Hôpital Croix-Rousse, 69317 Lyon, France

## Abstract

*Background and Aims:* DNA-based vaccination is a promising novel approach for immunotherapy of chronic Hepatitis B Virus (HBV) infection, however its efficacy needs to be improved. In this preclinical study we evaluated the therapeutic benefit of cytokine (IL-2, IFN-γ) genes co-delivery with DNA vaccine targeting viral proteins in the duck

* Correspondence concerning this article should be addressed to Lucyna Cova, Mailing address: UMR Inserm 1052, 151 cours Albert Thomas, 69003 Lyon, Phone: (33)4 72 68 19 81 ; Fax (33)4 72 68 19 71; E-mail address : lucyna.cova@inserm.fr

model of chronic HBV (DHBV) infection. DHBV is an avian hepadnavirus with replication strategies and genomic organization closely related to the human virus. Experimental DHBV inoculation of neonates always leads to the establishment of a chronic DHBV-carrier state, representing therefore a reference model for evaluation of novel therapeutic approaches on viral clearance, particularly on covalently closed circular (ccc) DNA elimination, viral minichromosome that is responsible for the chronicity of infection.

A colony of chronically infected Pekin ducks was established and treated by DNA vaccine encoding DHBV structural proteins (envelope, core) alone or co-delivered with either duck IL-2 (DuIL-2) or duck IFN-gamma (DuIFN-γ) gene. The impact of such therapy on viral replication was followed by monitoring of viremia, break of immune tolerance and intrahepatic viral liver DNA analysis.

*Results and Conclusion:* We showed that co-administration of cytokines genes (IL-2, IFN-γ) with DHBV DNA vaccine may represent a promising strategy able to enhance viral clearance from serum and liver of chronically infected carriers. IFN-γ gene co-delivery showed a particularly effective adjuvant activity able to enhance therapeutic efficacy of DNA vaccine targeting hepadnavirus proteins, leading to drastic inhibition or suppression of viral replication in about 50% of animals, which have only traces of cccDNA. Interestingly, the inhibition of viral replication was sustained, since no rebound of viral replication was observed during three months after immunotherapy cessation. However, some livers of apparently resolved animals were able to transmit DHBV infection to neonatal ducklings. No adverse effect was associated with co-imunnization with cytokine genes.

Collectively our results suggest a therapeutic benefit of DuIFN-γ gene co-delivery with DNA vaccine targeting viral proteins (envelope, core) for chronic hepatits B therapy. The efficacy in term of viral clearance was higher as compared with previous studies in which the therapeutic efficacy of DHBV DNA vaccine in combination or not with antiviral drugs (adefovir, lamivudine) has not exceeded 30%.

**Keywords:** DNA vaccine, Hepatitis B virus, duck HBV (DHBV), interferon-gamma, cccDNA

# Introduction

Hepatitis B virus (HBV) infection represents a major public health problem since it is associated with high mortality rates dues mainly to cirrhosis or hepatocellular carcinoma [1]. Considering the partial efficacy of current antiviral therapies (interferon-α, nucleoside analogs), there is an urgent need for developement more potent strategies [1]. Since chronic HBV infection is mostly characterized by varying degrees of T cells functional impairment and an unbalanced Th1 cytokine response (IFN-γ, IL-12, IL-2) [2], therefore DNA vaccination strategy able to stimulate both humoral and cellular arms of immune response particularly with Th1 profile, is considered of particular value for chronic hepatitis B therapy [3]. DNA-based vaccination targeting HBV proteins was shown to induce potent immune responses, leading to the suppression of viral replication in transgenic mice [4]. During first clinical trials in chronic HBV-carriers, DNA immunization efficacy showed a potential to stimulate immune responses although these responses were moderate and transient [5,6]. Improvement of DNA vaccine efficacy is therefore needed for future therapeutic applications in clinic. In this regard, it has been demonstrated, in naïve mice and woodchucks, that the co-administration of plasmids encoding cytokines such as interleukin-2 (IL-2) or interferon-gamma (IFN-γ) with DNA vaccine expressing hepadnavirus envelope or core proteins, enhanced humoral and cellular responses, with a reorientation to the Th1 profile [7-10]. However, evaluation of cytokine genes co-delivery with DNA vaccine in these studies was essentially limited to the increase of the immunogenicity or the prophylactic efficacy of such vaccine.

Data on cytokine genes co-administration on the enhancement of the therapeutic efficacy of DNA vaccine targeting HBV proteins are lacking. In addition, it remains largely unknown whether resolution of chronic HBV infection, is associated with a complete eradication of infection. In this view different studies showed the persistence of viral DNA in serum and PBMCs of patients and woodchucks longtime after resolution of infection and reactivation of viral replication following immunosupression or liver transplantation [11-22].

The duck HBV (DHBV) is a related avian hepadnavirus with replication strategies and genomic organization closely related to the human virus. Experimental DHBV inoculation to neonatal ducklings always leads to the establishment of a chronic DHBV-carrier state, representing therefore a reference model for evaluation of novel therapeutic approaches efficacy on viral clearance, particularly on cccDNA elimination [23]. We have previously reported that DNA vaccine-based immunotherapy of chronically infected ducks with plasmids encoding DHBV proteins (envelope, core) enhanced cccDNA clearance [24-26]. In addition, we recently demonstrated, in naïve ducks, that co-delivery of DuIFN-γ or DuIL-2 cytokine genes with DNA vaccine enhanced humoral response to DHBV large envelope protein (preS/S), which was highly neutralizing, particularly for duck groups co-immunized with Du-IFNγ encoding plasmid [27].

In the current study, we evaluated the therapeutic benefit of cytokine (DuIFN-γ, DuIL-2) genes co-delivery with DNA vaccine targeting viral proteins (envelope, core) in the chronic DHBV infection model. Moreover we searched for the persistence of replication-competent virus in the livers of animals that apparently resolved infection, presenting only traces of cccDNA undetectable by conventional methods.

# Materials and Methods

## Animals

DHBV-free Pekin ducklings (*Anas platyrhynchos*) were housed at the facilities of the National Veterinary School of Lyon (Marcy l'Etoile, France) in accordance with the guidelines for animal care.

## DNA Immunization

A colony of chronically infected ducks was established by intravenous inoculation of 3-day-old ducklings with pool of sera from ducklings that have been transfected with the cloned and sequenced DHBV [28], and quantified into virus genome equivalents (vge) as described previously [29]. At week 5 after hatching, DHBV carrier ducks were randomly assigned into

4 groups, which were immunized intramuscularly with a total of 450 μg of plasmid DNA (Table 1). pCI-preS/S, pCI-C, pCI-IL2 and pCI-IFNγ plasmids were constructed and verified for the expression of DHBV preS/S, core, DuIL-2 and DuIFN-γ respectively, as described previously [24,27,30,31]. Group 1 was immunized with empty plasmid pCI, whereas the remaining groups received immunizations with pCI-preS/S and pCI-C alone (group 2) or co-delivered with pCI-IL2 (group 3) or pCI-IFNγ plasmids (group 4) (Table 1). As summarized in Figure 1, booster doses were administered at weeks 8 and 11 post-hatching, followed by a delayed boost at week 20. After the last immunization, animals were followed during additional 14 weeks and were thereafter sacrificed. Necropsy liver samples were snap-frozen in liquid nitrogen and stored at-70°C before use for analysis described below.

**Table 1. Therapeutic protocol**

| Group | DNA immunization | Plasmid DNA constructs |
|---|---|---|
| 1 | pCI (controls) | 450 μg pCI |
| 2 | pCI-preS/S + pCI-C + pCI | 150 μg pCI-preS/S + 150 μg pCI-C + 150 μg pCI |
| 3 | pCI-preS/S + pCI-C + DuIL-2 | 150 μg pCI-preS/S + 150 μg pCI-C + 150 μg DuIL-2 |
| 4 | pCI-preS/S + pCI-C + DuIFN-γ | 150 μg pCI-preS/S + 150 μg pCI-C + 150 μg DuIFN-γ |

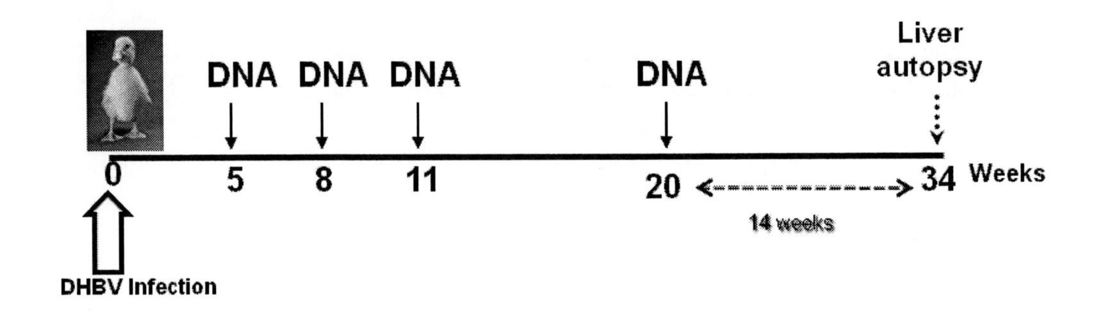


Figure 1. Design of therapeutic protocol. A colony of chronic DHBV-carrier ducks was established by inoculation of 3 days-old Pekin ducklings with DHBV positive inoculum (empty arrow). Black arrows indicate DNA injections at week 5, 8, 11 and a delayed boost at week 20.

## Analysis of Viral DNA

Viremia was assessed by quantitative detection of DHBV DNA in duck serum using dot blot hybridization assay described elsewhere [23]. Liver samples, collected during the twelve last weeks and at final necropsy, were submitted to two independent DNA extraction procedures for isolation of total viral DNA (protein bound) and for cccDNA (non protein bound) and analyzed by Southrn blotting as described previously [34,35]. For quantification of viral cccDNA by real time PCR total DNA was extracted from frozen autopsy liver samples, using the MasterPure DNA purification kit (Epicentre, Madison, USA) and quantified by spectrophotometry. Fifty ng of DNA were subjected to Plasmid-Safe™ ATP-dependent DNase (Epicentre, Madison, USA) digestion according to manufacturer's recommendations before PCR amplification as described (26). Determination of copy number of cccDNA was performed according to a standard curve based on agarose gel purified *Eco*RI-cleaved DHBV genome containing plasmid.

### Analysis of Infectivity of Liver Extracts Collected from Resolved Ducks

Small fragments (60 mg) of autopsy liver samples were homogenized in ice-cold PBS and clarified homogenates obtained from each of resolved ducks or from controls, were tested for presence of infectious virus by intravenous inoculation into 1-2 days old ducklings (4 ducklings/liver homogenate). Blood samples from inoculated ducklings were collected 2-5 times per week during 3-4 weeks of follow-up and were monitored for viremia.

### Histopathological and Immuno-Histochemical Analysis

Three μm thickness formalin-fixed liver tissue sections were stained with hematoxylin-eosin-safran stain and examined under code with a light microscope. The level of hepatocyte necrosis (acidopholic bodies), steatosis, portal tract and intralobular inflammation were assessed semi-quantitatively under code. DHBV preS antigen was detected by immunoperoxidase staining of liver sections using a previously characterized anti-preS murine monoclonal antibody and horseradish peroxydase-conjugated goat anti-mouse IgG as described previously [37]. All slides were counterstained with haematoxylin.

### Statistical Analysis

The efficiency of therapies on viremia, liver viral DNA, total viral clearance and seroconversion to anti-preS, were compared using the non parametric Mann-Whitney test to determine the significance between groups. Statistical significance was taken as $P<0.05$.

## Results

### Decrease in Viremia following Co-Immunization with Cytokine Expressing Plasmids

The analysis of serum DHBV DNA titers before the beginning of the treatment (week 0-5) showed no significant difference between the mean viremia levels of different groups. However, starting at week 21 *i.e.* after the last DNA boost and until the end of the follow-up (week 34), a marked decrease in mean viremia titers was observed for group 3 and 4 which reached significant levels ($P<0.05$) only for group 4 as compared with control group 1 and to group 2 receiving DHBV DNA vaccine alone.

Viral DNA was undetectable in the sera of one control duck (group 1) and number of animals from cytokine co-immunized groups: 2 of 8 co-immunized with IL-2 (group 2) and 4 of 7 ducks from DuIFN-γ co-immunized (group 4). DHBV DNA remained undetectable in the sera of all these 7 animals during the 3 months off DNA therapy (week 20-34) suggesting a sustained effect of DNA vaccine-based therapy.

## Enhancement of DHBV DNA clearance by Co-Immunization with IFN-γ expressing Plasmid

To search whether undetectable viremia was associated with intrahepatic viral DNA clearance we analyzed first by Southern blotting the DNA extracted from necropsy liver samples from all 30 animals included in this study. Analysis of intrahepatic DHBV DNA from empty vector-treated controls (group 1) showed the presence of viral DNA replicative forms for all, but one duck presenting undetectable DHBV DNA, consistently with the absence of detectable viremia for this animal. Interestingly, 2 out of 8 (25%) and 4 out of 7 (57%) ducks co-immunized with pCI-IL-2 or pCI-IFNγ plasmids, respectively eliminated all intrahepatic DHBV DNA intermediates, including cccDNA (data not shown). The absence of detectable cccDNA in these liver samples was further confirmed by Southern blot analysis following an independent non-protein bound cccDNA extraction (data not shown).

Next, we searched for the persistence of residual intrahepatic viral DNA, undetectable by conventional methods (such as dot or Southern blot analysis) by a real-time PCR assay specific for cccDHBV DNA in these liver samples. Our results revealed that all 7 resolved animals, which had undetectable liver DHBV DNA replicative forms in Southern blotting, presented traces of cccDNA in real-time PCR analysis being about 3-5 logs lower as compared to controls presenting ongoing DHBV replication.

## Virological Analysis and Infectivity of Liver Samples Collected from Resolved Animals

To better understand whether the traces of DHBV DNA persisting in the liver of apparently resolved animals were associated with the presence of infectious virions, we set up an *in vivo* infectivity assay, based on the inoculation of neonatal ducklings, highly susceptible to DHBV infection, with clarified necropsy liver extracts prepared from apparently resolved and control animals. Most of control animals that received empty vector immunization, exhibited high serum viral load throughout the study and absence of detectable anti-preS response. This profile was associated with a massive expression of DHBV preS protein in virtually all hepatocytes. However, different degrees of viral clearance and infection transmission were observed in livers of apparently resolved animals that cleared viremia and presented only traces of cccDNA. Transmission of infection was observed for resolved animals exhibiting low level transient seroconversion, and presenting DHBV preS expression restricted to few hepatocytes or undetectable by immunostaining, indicating that extremely low residual viral expression or replication may persist in the livers of these animals. Whereas, absence of infection transmission, associated with undetectable DHBV preS expression and correlated or not with seroconversion to more sustained anti-preS response, was observed for 3 out of 4 resolved ducks co-immunized with DuIFN-γ plasmid (group 4), suggesting a more complete stage of viral clearance in this group.

### Adverse Effects of Therapy

Histological analysis of duck liver samples showed an absence of severe lesions, necrosis or fulminant hepatitis for all ducks. Occasional presence of amyloidosis, known to occur in adult Pekin ducks [38], was observed regardless of treatment group. In addition, inflammatory infiltrates observed in portal tracts of treated ducks were comparable to controls. Several deaths have occurred during the follow up, although they were accidental and their incidence rate was similar between groups receiving DNA vaccine alone or combined with cytokines expressing plasmids. These data indicates the absence of notable side effects for the treatment used by us under the described conditions.

## Conclusion

We report here that co-delivery of cytokine encoding plasmids, especially IFN-γ expressing plasmid considerably enhanced therapeutic efficacy of DNA vaccination in DHBV-carriers, suggesting the benefit of this approach for chronic hepatitis B treatment. The comparison of IL-2 with IFN-γ gene co-delivery indicated that co-immunization of DHBV-carriers with DuIFN-γ encoding plasmid was particularly effective leading to a significant drop in viremia ($P<0.05$) starting at week 21 p.i. and until the end of the follow-up, as compared to controls and duck group treated with DHBV DNA vaccine alone. In addition, no rebound in viremia was observed in animals 3 months after immunotherapy cessation, indicating sustained effect of this approach. In addition, viral DNA elimination in serum was correlated with the clearance of intrahepatic viral DNA replicative forms, including cccDNA, in more than 50% of animals. Whereas, co-immunization with DuIL-2 encoding plasmid was less pronounced, with only 25% of treated animals that eliminated serum and liver DHBV DNA. It is to note that viral clearance was not always linked with the appearance of high anti-preS antibody titers. This could be related to the presence of immune complexes that can not be detected by our ELISA test, and/or to the fact that T cell immune response induced by preS/S and core immunization may play an essential role in viral clearance. We were unable to investigate the impact of duck cytokines co-delivery on the cellular response induction since tools for its analysis in the duck model are still under development.

The biological significance of residual viral DNA, frequently detected following resolution of hepadnaviral infection, remains an open question. Detection of intrahepatic viral DNA, including the cccDNA replicative form has been reported earlier in patients and woodchucks with resolved infection [11,14,15,18] and in the duck model [36]. Reactivation of HBV infection occurred in some patients following immunosuppressive treatment or liver transplantation or haemodialysis and was demonstrated recently in the woodchuck model following immunosuppressive treatment of certain resolved animals [20-22]. However, in the duck model, it was reported that persistent cccDNA, detected nine months following apparent resolution of transient DHBV infection, was not associated with the presence of ongoing infection and may be present only in a steady state non replicative form [36]. Thus, data concerning the potential of traces of viral DNA to reactivate the infection are scarce and contradictory. Using an *in vivo* infectivity test we searched for the persistence of viral infectivity in the liver of ducks that apparently resolved DHBV infection, presenting traces of residual viral DNA detectable by real-time PCR only. We observed the establishment of productive infection in ducklings inoculated with liver extracts of 3 out of 7 ducks that had undetectable serum DHBV DNA during last 14 weeks, indicating the persistence of replication-competent virus in their livers extracts. The fact that liver homogenates from 3 of 4 ducks belonging to IFN-γ plasmid immunized group did not transmitted DHBV infection could be explained by a more complete stage of clearance that may be due to more potent cellular immune responses.

The safety of DNA-based immunization against DHBV large envelope protein and DHBV core has been previously demonstrated [24,25]. In the present study, no particular adverse effects were associated with co-immunization with cytokine genes. It was previously showed that intravenous injections of IL-2 protein to chronic HBV carriers showed severe toxicities [41]. The discrepancy in the side effects between our DNA vaccine and previous IL-2 protein administration may be due to the lowest *in vivo* level of the pro-inflammatory cytokine encoded by DNA vaccine, compared to the protein administration.

In conclusion, our study in DHBV infection model showed that co-administration of Th1 cytokines (IL-2, IFN-γ) with HBV DNA vaccine may represent a promising strategy able to enhance viral clearance from sera and livers of chronically infected carriers. In particular IFN-γ gene co-delivery showed a particularly effective adjuvant activity able to enhance therapeutic efficacy of DNA vaccine targeting hepadnavirus proteins, with a sustained and drastic inhibition or suppression of viral replication in more than 50% of chronic DHBV carrier animals. This is of particular interest, since in previous studies the therapeutic efficacy of DHBV DNA vaccine in combination or not with antiviral drugs (adefovir, lamivudine) has not exceeded 30% [25,26].

# Acknowledgments

This study was supported by a grant from French National Agency of Research on Aids and Hepatitis B and C (ANRS). Fadi SAADE was the recipient of a fellowship from ANRS and Ghada Khawaja received a fellowship from the Research Cluster of Infectiology (Region Rhône-Alpes).

# References

[1] Lavanchy, D. Hepatitis B virus epidemiology, disease burden, treatment, and current and emerging prevention and control measures. *J. Viral. Hepat,* 2004, 11, 97-107.

[2] Rehermann, B. Chronic infections with hepatotropic viruses: mechanisms of impairment of cellular immune responses. *Semin. Liver Dis*., 2007, 27, 152-160.

[3] Donnelly, JJ; Ulmer, JB; Shiver, JW; Liu, MA. DNA vaccines. *Annu. Rev. Immunol.,* 1997, 15, 617-648.

[4] Mancini, M; Hadchouel, M; Davis, HL; Whalen, RG; Tiollais, P; Michel, ML. DNA-mediated immunization in a transgenic mouse model of the hepatitis B surface antigen chronic carrier state. *Proc. Natl. Acad. Sci. USA.,* 1996, 93, 12496-12501.

[5] Mancini-Bourgine, M; Fontaine, H; Scott-Algara, D; Pol, S; Brechot, C; Michel, ML. Induction or expansion of T-cell responses by a hepatitis B DNA vaccine administered to chronic HBV carriers. *Hepatology*, 2004, 40, 874-882.

[6] Yang, SH; Lee, CG; Park, SH; Im, SJ; Kim, YM; Son, JM; Wang, JS; Yoon, SK; Song, MK; Ambrozaitis, A; Kharchenko, N; Yun, YD; Kim, CM; Kim, CY; Lee, SH; Kim, BM; Kim, WB; Sung, YC. Correlation of antiviral T-cell responses with suppression of viral rebound in chronic hepatitis B carriers: a proof-of-concept study. *Gene Ther.,* 2006,13, 1110-1117.

[7] Wang, J; Gujar, SA; Cova, L; Michalak, TI. Bicistronic woodchuck hepatitis virus core and gamma interferon DNA vaccine can protect from hepatitis but does not elicit sterilizing antiviral immunity. *J. Virol.,* 2007, 81, 903-916.

[8] Chow, YH; Chiang, BL; Lee, YL; Chi, WK; Lin, WC; Chen, YT; Tao, MH. Development of Th1 and Th2 populations and the nature of immune responses to hepatitis B virus DNA vaccines can be modulated by codelivery of various cytokine genes. *J. Immunol.,* 1998, 160, 1320-1329.

[9] Geissler, M; Schirmbeck, R; Reimann, J; Blum, HE; Wands, JR. Cytokine and hepatitis B virus DNA co-immunizations enhance cellular and humoral immune responses to the middle but not to the large hepatitis B virus surface antigen in mice. *Hepatology,* 1998, 28, 202-210.

[10] Siegel, F; Lu, M; Roggendorf, M. Coadministration of gamma interferon with DNA vaccine expressing woodchuck hepatitis virus (WHV) core antigen enhances the specific immune response and protects against WHV infection. *J. Virol*., 2001, 75, 5036-5042.

[11] Michalak, TI; Pasquinelli, C; Guilhot, S; Chisari, FV. Hepatitis B virus persistence after recovery from acute viral hepatitis. *J. Clin. Invest*., 1994, 94, 907.

[12] Penna, A; Artini, M; Cavalli, A; Levrero, M; Bertoletti, A; Pilli, M; Chisari, FV; Rehermann, B; Del Prete, G; Fiaccadori, F; Ferrari, C. Long-lasting memory T cell responses following self-limited acute hepatitis B. *J. Clin. Invest.,* 1996, 98, 1185-1194.

[13] Rehermann, B; Ferrari, C; Pasquinelli, C; Chisari, FV. The hepatitis B virus persists for decades after patients' recovery from acute viral hepatitis despite active maintenance of a cytotoxic T-lymphocyte response. *Nat. Med.* 1996, 2, 1104-1108.

[14] Marusawa, H; Uemoto, S; Hijikata, M; Ueda, Y; Tanaka, K; Shimotohno, K; Chiba, T. Latent hepatitis B virus infection in healthy individuals with antibodies to hepatitis B core antigen. *Hepatology*, 2000, 31, 488-495.

[15] Yuki, N; Nagaoka, T; Yamashiro, M; Mochizuki, K; Kaneko, A; Yamamoto, K; Omura, M; Hikiji, K; Kato, M. Long-term histologic and virologic outcomes of acute self-limited hepatitis B. *Hepatology*, 2003, 37, 1172-1179.

[16] Coffin, CS; Michalak, TI. Persistence of infectious hepadnavirus in the offspring of woodchuck mothers recovered from viral hepatitis. *J. Clin. Invest.,* 1999, 104, 203-212.

[17] Michalak, TI; Pardoe, IU; Coffin, CS; Churchill, ND; Freake, DS; Smith, P; Trelegan, CL. Occult lifelong persistence of infectious hepadnavirus and residual liver inflammation in woodchucks convalescent from acute viral hepatitis. *Hepatology*, 1999, 29, 928-938.

[18] Menne, S; Cote, PJ; Butler, SD; Toshkov, IA; Gerin, JL; Tennant, BC. Immuno suppression reactivates viral replication long after resolution of woodchuck hepatitis virus infection. *Hepatology,* 2007, 45, 614-622.

[19] Michalak, TI; Mulrooney, PM; Coffin, CS. Low doses of hepadnavirus induce infection of the lymphatic system that does not engage the liver. *J. Virol.,* 2004, 78, 1730-1738.

[20] Dickson, RC; Everhart, JE; Lake, JR; Wei, Y; Seaberg, EC; Wiesner, RH; Zetterman, RK; Pruett, TL; Ishitani, MB; Hoofnagle, JH. Transmission of hepatitis B by trans plantation of livers from donors positive for antibody to hepatitis B core antigen. The National Institute of Diabetes and Digestive and Kidney Diseases Liver Transplantation Database. *Gastroenterology*, 1997, 113, 1668-1674.

[21] Lok, AS; Liang, RH; Chiu, EK; Wong, KL; Chan, TK; Todd, D. Reactivation of hepatitis B virus replication in patients receiving cytotoxic therapy. Report of a prospective study. *Gastroenterology*, 1991, 100, 182-188.

[22] Markovic, S; Drozina, G; Vovk, M; Fidler-Jenko, M. Reactivation of hepatitis B but not hepatitis C in patients with malignant lymphoma and immunosuppressive therapy. A prospective study in 305 patients. *Hepatogastroenterology* 1999, 46, 2925-2930.

[23] Cova L, Zoulim F. Duck hepatitis B virus model in the study of hepatitis B virus. *Methods Mol. Med.,* 2004, 96, 261-268.

[24] Rollier, C; Sunyach, C; Barraud, L; Madani, N; Jamard, C; Trepo, C; Cova, L. Protective and therapeutic effect of DNA-based immunization against hepadnavirus large envelope protein. *Gastroenterology*, 1999, 116, 658-665.

[25] Le Guerhier, F; Thermet, A; Guerret, S; Chevallier, M; Jamard, C; Gibbs, CS; Trepo, C; Cova, L; Zoulim F. Antiviral effect of adefovir in combination with a DNA vaccine in the duck hepatitis B virus infection model. *J. Hepatol.,* 2003, 38, 328-334.

[26] Thermet, A; Buronfosse, T; Werle-Lapostolle, B; Chevallier, M; Pradat, P; Trepo, C; Zoulim, F; Cova, L. DNA vaccination in combination or not with lamivudine treatment breaks humoral immune tolerance and enhances cccDNA clearance in the duck model of chronic hepatitis B virus infection. *J. Gen. Virol.,* 2008, 89, 1192-1201.

[27] Saade, F; Buronfosse, T; Pradat, P; Abdul, F; Cova, L. Enhancement of neutralizing humoral response of DNA vaccine against duck hepatitis B virus envelope protein by co-delivery of cytokine genes. *Vaccine,* 2008, 26, 5159-5164.

[28] Mandart, E; Kay, A; Galibert, F. Nucleotide sequence of a cloned duck hepatitis B virus genome: comparison with woodchuck and human hepatitis B virus sequences. *J. Virol.,* 1984, 49, 782-792.

[29] Borel, C; Sunyach, C; Hantz, O; Trepo, C; Kay, A. Phosphorylation of DHBV pre-S: identification of the major site of phosphorylation and effects of mutations on the virus life cycle. *Virology*, 1998, 242, 90-98.

[30] Thermet, A; Robaczewska, M; Rollier, C; Hantz, O; Trepo, C; Deleage, G; Cova, L. Identification of antigenic regions of duck hepatitis B virus core protein with antibodies elicited by DNA immunization and chronic infection. *J. Virol.,* 2004, 78, 1945-1953.

[31] Narayan, R; Buronfosse, T; Schultz, U; Chevallier-Gueyron, P; Guerret, S; Chevallier, M; Saade, F; Ndeboko, B; Trepo, C; Zoulim, F; Cova, L. Rise in gamma interferon expression during resolution of duck hepatitis B virus infection. *J. Gen. Virol.,* 2006, 87, 3225-3232.

[32] Lambert, V; Chassot, S; Kay, A; Trepo, C; Cova, L. *In vivo* neutralization of duck hepatitis B virus by antibodies specific to the N-terminal portion of pre-S protein. *Virology*, 1991, 185, 446-450.

[33] Rollier, C; Charollois, C; Jamard, C; Trepo, C; Cova, L. Early life humoral response of ducks to DNA immunization against hepadnavirus large envelope protein. *Vaccine,* 2000, 18, 3091-3096.

[34] Jilbert, AR; Wu, TT; England, JM; Hall, PM; Carp, NZ; O'Connell, AP; Mason, WS. Rapid resolution of duck hepatitis B virus infections occurs after massive hepatocellular involvement. *J. Virol.,* 1992, 66, 1377-1388.

[35] Le Guerhier, F; Pichoud, C; Guerret, S; Chevallier, M; Jamard, C; Hantz, O; Li, XY; Chen, SH; King, I; Trepo, C; Cheng, YC; Zoulim, F. Characterization of the antiviral effect of 2',3'-dideoxy-2', 3'-didehydro-beta-L-5-fluorocytidine in the duck hepatitis B virus infection model. *Antimicrob. Agents Chemother.,* 2000, 44, 111-122.

[36] Le Mire, MF; Miller, DS; Foster, WK; Burrell, CJ; Jilbert, AR. Covalently closed circular DNA is the predominant form of duck hepatitis B virus DNA that persists following transient infection. *J. Virol.*, 2005, 79, 12242-12252.

[37] Barraud, L; Guerret, S; Chevallier, M; Borel, C; Jamard, C; Trepo, C; Wild, CP; Cova, L. Enhanced duck hepatitis B virus gene expression following aflatoxin B1 exposure. *Hepatology*, 1999, 29, 1317-1323.

[38] Cova, L; Wild, CP; Mehrotra, R; Turusov, V; Shirai, T; Lambert, V; Jacquet, C; Tomatis, L; Trepo, C; Montesano, R. Contribution of aflatoxin B1 and hepatitis B virus infection in the induction of liver tumors in ducks. *Cancer Res.,* 1990, 50, 2156-2163.

[39] Gurunathan, S; Klinman, DM; Seder, RA. DNA vaccines: immunology, application, and optimization*. *Annu .Rev. Immunol.*, 2000, 18, 927-974.

[40] Thermet, A; Rollier, C; Zoulim, F; Trepo, C; Cova, L. Progress in DNA vaccine for prophylaxis and therapy of hepatitis B. *Vaccine*, 2003, 21, 659-662.

[41] Bruch, HR; Korn, A; Klein, H; Markus, R; Malmus, K; Baumgarten, R; Muller, R. Treatment of chronic hepatitis B with interferon alpha-2b and interleukin-2. *J. Hepatol.,* 1993, 17, S52-55.

In: DNA Vaccines: Types, Advantages and Limitations ISBN 978-1-61324-444-9
Editors: E. C. Donnelly and A. M. Dixon 

*Chapter VI*

# The Role of DNA-IL-12 Vaccination in Eosinophilic Inflammation: A Review

***A. Malheiro,***[*1,4] ***L. de Paula,***[1,2] ***O. A. Martins-Filho,***[3] ***F. F. Anibal,***[5]
***A. Teixeira-Carvalho,***[3] ***R. Alle-Marie***[4] ***and L. H. Faccioli***[6]
[1]Federal University of Amazonas
[2]Alfredo da Matta Foundation
[3]Reneé Rachou Institute – Fiocruz
[4]Amazon Hematology and Hematherapy Foundation
[5]Universidade Federal de São Carlos
[6]College of Pharmaceutical Sciences of Ribeirão Preto-Universidade de São Paulo, Brazil

## Abstract

DNA-based vaccines have garnered attention for their potential as alternative treatments for established diseases. Eosinophilia is a key inflammatory feature of allergic pathologies including asthma. This pathology is characterized by airway obstruction, airway hyper-responsiveness and lung tissue remodeling. IL-12 is a potent inducer of Th1 cellular immune responses, but maintenance of the Th1 immune response is necessary for increased IFN-γ production. Th1 cytokines such IL-12 and IFN-γ have been shown to reduce lung allergic responses. We have previously reported that DNA-IL-12 vaccination leads to a persistent reduction in blood/bronchoalveolar eosinophilia following *Toxocara canis* infection. Prominent type-1 immune response has been identified as the hallmark of *T. canis* infection following DNA-IL-12 vaccination. The type-1 polarized immune-logical profile in DNA-IL-12-vaccinated animals is characterized by a high IFN-γ/IL-4 ratio and low levels of IgG1, with subsequent high IgG2a/IgG1 ratio. The persistent airway hyper-responsiveness observed in DNA-IL-12-vaccinated animals demonstrates that the airway constriction observed involves immunological mediators other than those blocked by DNA-IL-12. The data discussed here provide evidence of the potential utility

[*] Address for correspondence: Dra. Adriana Malheiro, Av. Constantino Nery 4397 69050-002 Manaus, AM, Brazil. Tel: 55 92 36550113, email: malheiroadriana@yahoo.com.br.

of DNA-IL12 vaccine for the development of a new treatment for eosinophilic inflammation. The purpose of this study is to review DNA vaccination, a novel strategy of DNA administration, as well as the role of DNA-IL-12 vaccination in eosinophilic inflammation.

# Basic DNA Vaccine Concepts

The DNA or third-generation vaccine was developed within the last two decades. The technology involves the association of genes or genetic fragments, which code potentially immunodominant antigens with known DNA vectors as plasmids [1,2]. The plasmids used in DNA vaccines possess DNA sequences necessary for selection and replication in bacteria; special promoters for transcription processes; genes that confer antibiotic resistance, for the selection of transformed bacteria; and specific sequences that allow for gene expression in eukaryotic or prokaryotic cells.

This vaccination strategy has been used as prevention or infectious disease therapy in different experimental models. Ulmer and collaborators in (1993) reported that intramuscular injection of a plasmid containing the gene that codes for influenza virus nucleoprotein could be used for immunization of mice against this virus, a fact that led to the development of new studies in this area [3]. These studies demonstrated that the great advantage of using vaccines based on plasmid DNA containing immunogenic protein genes is that plasmid DNA supplies necessary genetic information, so that the host organism produces the antigen using the cellular machinery, without the undesirable effect of introducing the pathogenic infectious agent or vaccines containing protein subunits or adjuvant [4]. These vaccines present advantages such as induction of both types of immune responses (cellular and humoral), offering the possibility of combining antigens from one or many pathogens or even co-administration with immunomodulators. Furthermore, these constructs are easy to produce and present good stability.

The first step in genetic immunization is in vivo transfection of plasmid. After intramuscular administration, the purified DNA can be incorporated in myocytes or mononuclear cells such as macrophages or dendritic cells, but the immune response stimulation involves cells derived from the bone marrow that present the antigen [5-7]. The cellular transfection mechanism involved with the use of purified DNA is not completely understood; however, it is known that the purified DNA reaches the cellular nucleus, where it remains in an episomal form; that is, without incorporating into the host cell's genome. The metabolic pathways of the host cell are used for transcription of genes, present in the DNA vaccine, as messenger RNA, which are translated into the protein antigen expressed by the infectious agent [5]. Thus, the process that occurs after production of the antigenic protein is very similar to that observed in viral replication, which results in the development of a cellular and humoral immune response [3,7].

While intracellular production of the antigen can stimulate the immune response, the vector itself can contribute to the effectiveness of this response. The plasmid vectors can present immunostimulatory properties, acting as adjuvant, increasing the magnitude of the immune response and influencing the standard of the cellular response. These immunostimulatory properties of the bacterial DNA are due to structural differences between DNA of eukaryotes and the presence of short base sequences that are characteristic of bacterial DNA. These immunostimulatory sequences, or CpG domains, have the general structure of two purines in the 5' position, an unmethylated dinucleotide CpG and two pyrimidines in the 3' position. The CpG motif functions as a danger signal, activating innate immunity, by binding to the "Toll-like"-9 receptor present on the surface of some cells [8-10]. Its immunostimulatory properties induce the production of cytokines such as IFN-αβγ, IL-12, TNF-α and IL-6, B cell proliferation and antibody synthesis [11-14]. Krieg and collaborators [1998] demonstrated that vaccination with DNA can induce changes in the standard immunological response, due to the adjuvant effect of the plasmid. It was also demonstrated that the CpG motif is capable of modulating allergic inflammation [15].

Various models using the DNA vaccine strategy have been described. For example, one model involved use of the DNA vaccine strategy against Gram-positive or Gram-negative bacteria [16]. In addition, Weiss and collaborators (1998) demonstrated that plasmids expressing GM-CSF increased immunogenicity against the parasite Plasmodium yoelii, in a murine model of malaria. Data from the literature demonstrate that the DNA vaccine was efficient in the treatment of tuberculosis. DNA that codes for IL-12, or both IL-12 and HSP-65, significantly reduced the population of Mycobacterium tuberculosis in different organs, for a sustained period [1].

The stimulation of human cells with the hsp-65 DNA vaccine has been studied. Recent works demonstrate that macrophages and dendritic cells are stimulated and respond in different ways. The macrophages respond by increasing the production of IL-6 and IL-10, with limited production of TNF-α. The dendritic cells produce high levels of IL-12 and low levels of TNF-α, IL-6 and IL-10 [17]. In this way, the targeting of DNA to specific cells is fundamental to the vaccination strategy.

## Strategies for DNA Vaccine Administration

Many advances have occurred over the last few decades in the development of peptide, pharmacological and DNA-release strategies. The traditional vaccination protocols with naked DNA require large quantities of DNA, since a reinforcing dose is needed. One possibility that is widely studied is the targeting of DNA vaccines to cells of interest. It was demonstrated that the effectiveness of gene vaccination is increased when the DNA is directed to the dendritic cells, resulting in a greater quantity of TCD4+ and CD8+ IFN-γ-specific producing cells [18].

The role of carrier particles in delivering DNA vaccines has been studied widely. The size of particles must be adjusted so that the DNA contained inside of these can be directed to the antigen-presenting cells, inducing a specific immune response, as macrophages are capable of phagocytosing particles ranging from 0.5 to 3 μm in diameter [19].

The majority of the protocols used for gene transfer in somatic animal cells involve *in vitro* transformation of these cells. However, these cells can also be transfected *in vivo*, using the biobalistic method, which is simple and effective. The biobalistic or "gene gun" involves the transfection of DNA attached to gold particles (0.6 to 2 μm in diameter). To distribute these particles throughout the target tissue, they are submitted to a shock wave generated by chemical explosion, helium discharge under high pressure or compressed air discharge [20]. The particles continue to move until reaching the target tissue. Barry and Johnston (1997) showed that intramuscular injection requires plasmid DNA content almost 100 times greater than the level required for the "gene gun." This difference in effectiveness can be attributed to the fact that intramuscular injection places the plasmid in an extracellular environment, where the majority of the DNA is quickly degraded by nucleases [21]. In contrast, the "gene gun" liberates the plasmid inside the cells, thus preventing its degradation.

Various studies have used the "gene gun" as a vaccination strategy against different types of infectious agents, such as *Mycoplasma pulmonis* [22], *Plasmodium spp.* [23] and the Ebola virus [24]. Pertmer and collaborators (1995) have reported that immunization with the "gene un" induces levels of antibody production similar to those observed when the vaccine is administered through the intramuscular route [25]. Using the same vaccine preparation, Feltquate and collaborators (1997) have demonstrated that the type of immune response induced by the "gene gun" immunization is predominantly Th2, differing from the response induced by intramuscular route immunization, which is predominantly Th1 [26]. However, studies suggest that the manner by which DNA is presented to the cells of the organism can influence its capture and consequently immune response induction.

Another alternative for DNA presentation is the use of microspheres containing DNA. These microspheres have been used in various experimental models with proven effectiveness in the induction of a protective and persistent immune response. Moreover, the use of polylactic acid co-glycolic acid (PLGA) microspheres containing phospholipase A2, the principal allergen of bee and wasp venom, resulted in induction of tolerance in a murine model, with reduced IgE levels and decreased T-cell hypo-responsiveness [27]. Nevertheless, there still are no published works that focus on the importance of vaccination with PLGA microspheres in the protection against respiratory diseases such as asthma.

The microsphere technology used for DNA vaccines allowed for the development of a new alternative in the field of gene vaccination: combining small microspheres (and subsequent rapid consumption of the DNA) with larger microspheres, which persist for more time, thus delaying the liberation of DNA. In this way, the plasmid containing the DNA can be consumed at two time-points, rapidly from the smaller microspheres and at a second stage, occurring days later, when the constituent microsphere material will have been consumed. This strategy has shown good results in mice and guinea pigs immunized with DNA vaccine tuberculosis encapsulated in PLGA microspheres [28].

Various alternative protocols of priming-boosting with protein have also been used in DNA vaccine studies. In these protocols, the immunization is conferred by the DNA and the protein assists in the induction of a more efficient immune response. The priming-boosting model resulted in increased numbers of cytotoxic T cells and IFN-$\gamma$ secretory cells against the hepatitis C virus when compared with gene vaccination without the protein [29]. A greater protective effect was also demonstrated when priming-boosting used DNA protein against *Plasmodium knowlesi* infection in rhesus monkey [30].

Although the epidermis acts as an important barrier against invading microorganisms, its thick stratum corneum also hinders the passage of pharmacological compounds. For this reason, recent research has focused on the importance of microneedles that release substances without stimulating the nerve cells that trigger painful sensations. High-molecular weight antigens are freed more slowly in the skin, and the use of hollow microneedles increases the effectiveness of release of these compounds [31]. This strategy will be able to help patients who present with phobias to traditional injection. However, no published study has examined the use of microneedles in asthma or other models of allergic disease.

Another powerful strategy for DNA vaccination involves the use of liposomes. It has been demonstrated that cationic liposomes can be used for gene release with the same protective efficiency as observed for intramuscular DNA inoculation [32]. The first drug to be commercialized with this technology was the synthetic pulmonary surfactant Alveofact (Dr. Karl Thomae GmbH, Germany) used for the treatment of respiratory distress syndrome (RDS). Based on the action of liposomes in the lungs, this controlled release alternative could be beneficial in the context of asthma and other respiratory illnesses [33].

## Structure and Function of Interleukin-12

Interleukin-12 (70 IL-12) is a 70-kDa heterodimer formed by two covalently bonded glycosylated chains, of approximately 40 (p40) and 35 (p35) kDa [34-37]. This cytokine is produced by phagocytes, B lymphocytes and other antigen-presenting cells in response to bacteria, bacteria products and intracellular parasites [38]. It acts in a paracrine manner, inducing growth and activation of NK cells and increasing the lytic potential of these cells and CD8+ T lymphocytes [36;39;40]. The production of IL-12 by phagocytes favors a standard T "helper" 1 response, involving the production of IL-2 and IFN-γ [41-46] and other cytokines, such as TNF-α, GM-CSF, M-CSF, IL-3, and IL-8 [34,37,39].

Studies have demonstrated that "naive" T cells stimulated with IL-12 differentiate into CD4+ T cells of the T "helper" 1 type, which produce high concentrations of IFN-γ and low concentrations of IL-4 [47]. Marshall and collaborators (1995) reported that IL-12 inhibited the production of IL-4 and IL-10 and increased the synthesis of IFN-γ and IL-2 in T cells, after stimulation with the allergen [48]. These authors suggested that the administration of IL-12 could be effective in the therapy of patients with illnesses characterized by high levels of IL-4 (for example, allergy or leprosy), by inhibiting the synthesis of this cytokine and increasing the production of IFN-γ [48]. On the other hand, other authors demonstrated that IL-12 increases the production of IFN-γ without inhibiting IL-4 production. These authors suggested that IL-12 can have a limited effect on IL-4 production [45].

## IL-12 Therapy

Until now, no satisfactory therapeutic intervention has existed for the treatment of allergic diseases and asthma. However, IL-12 has been the object of various studies and is currently described as a cytokine that inhibits allergic inflammatory reactions (involving the aerial pathway) and asthma [49]. Lee and collaborators (2001) evaluated the effect of oral IL-12 treatment in a murine anaphylactic shock model and demonstrated a therapeutic effect, as well as the alleviation of symptoms and diminished release of histamine in the plasma [49]. Marinaro and collaborators (1997) have demonstrated that IL-12 treatment and sensitization with antigen reduces the Th2 response and increases the Th1 response, without affecting systemic IgA or IL-12 secretion levels [50].

Other authors have reported the participation of IL-12 in inhibition of the eosinophilic inflammatory process. Gavett and collaborators (1995) have demonstrated that treatment of experimental animals with IL-12 inhibits the airway hyperresponsiveness induced by antigen administration [51]. Subsequently, other authors using the chimeric Ova/IL-12 protein reported its efficacy in directing the immune response to Th1 and in inhibiting the production of specific IgE [52]. Some works have demonstrated that mice immune to Ova and stimulated with this protein present reduced airway hyperreactivity and inhibition of IgE antibody synthesis [53] when treated with IL-12 [51,53]. Moreover, IL-12 used as adjuvant in the vaccination with *S. mansoni* modifies the standard Th2 response to Th1 and inhibits the formation of granuloma during the reinfection [54]. In addition, animals infected with *N. brasilienses* and treated with IL-12 present reduced Th2 cytokine expression and increased IFN-γ expression [55]. The change in the cytokine pattern observed with this model is inhibited IgE production [55]. Similarly, Stampfli and collaborators (1999) have demonstrated that IL-12 gene transfer to the airways of mice is efficient in inhibiting eosinophilic inflammation, hyperresponsivity and IgE production, as well as modifying the standard immunological Th2 response to Th1 [56].

IL-12 has been used in other animal models, for example in the study of Acquired Immunodeficiency Syndrome [57], gene therapy for cancer [58] and the development of a vaccine against *Leishmania spp.* [59]. In mice infected with *L. monocytogenes* [60] or *Toxoplasma gondii* bacteria [61], IL-12 is produced rapidly, inducing the production of IFN-γ by NK cells and T cells.

## Eosinophilia and Asthma

Eosinophils account for 1-5% of peripheral leukocytes. Eosinophils are produced in the bone marrow from pluripotent stem cells; these cells differentiate into precursor cells with properties common to eosinophils and basophils [62]. The differentiation and proliferation of eosinophils are regulated by factors such as GM-CSF, IL-3 and IL-5 [63-65]. However, IL-5 alone is sufficient for the final differentiation of eosinophils [65-67]. These cells are multifunctional leukocytes that release large amounts of immunomodulatory compounds, including potent inducers of the inflammatory and immune responses seen in asthma, rhinitis, eczema, autoimmune diseases and cancer.

Asthma is a disease characterized by airway inflammation, variable airflow obstruction and airway hyperresponsiveness (AHR); infiltration of eosinophils, mast cells and CD4+ lymphocytes; and airway remodeling. Excessive production of interleukin (IL) -4, -5 and -13 by T helper type 2 (Th2) cells has been implicated in the pathogenesis of allergic asthma [55,68]. IL-5 mobilizes eosinophils from the bone marrow pool, whereas chemokines such as eotaxin-1 recruit eosinophils to the upper airways [65,69-71]. Among the mediators involved in asthma, cytokines play an important role in the immune response. The allergic inflammation in asthma is a characteristic inflammatory response, which involves a reaction to a harmless substance, an allergen. This leads to the production of specific IgE and then to the aggregation of high-affinity IgE (FceRI) receptors expressed on the surfaces of mast cells and eosinophils [72]. Some authors have related the degree of eosinophilia in the airways with bronchial hyperresponsiveness and clinical asthma symptoms [55,73,74]. The profile of the Th2 response plays an important role in the development and maintenance of eosinophilia [65,75-77,78]. However, the cytokine profile of the Th1 immune response can inhibit eosinophilic inflammation by suppressing Th2 development [79]. IL-12 acts mainly to regulate the Th2 response and has been proposed as a tool that can be used to treat allergies [45,48,80]. It is suggested that IL-12 is produced in the early phase of allergic inflammation to modulate the immune response to Th1. This results in eosinophilia control and airway hyperresponsiveness [80]. Our group and other authors have evaluated a therapeutic approach using an animal model of the eosinophilic inflammatory response. These studies evaluated: (i) key development areas for the use of therapeutic cytokines, including monoclonal antibodies against Th2 cytokines [81]; (ii) manipulation of the immune response directly or indirectly, including DNA administration of IL-12 [80].

## Impact of pcDNA-IL-12 Administration on Eosinophilia and Pulmonary Hyperreactivity in a Murine Model

Many experimental models have been used to elucidate the eosinophilic inflammatory process. Such models involve induction by immunization with ovalbumin [82,83]; *S. mansoni* [49] and *T. canis* [65], among others. We used the Visceral Larval Migration Syndrome model (VLMS) and toxocariasis to study eosinophilic inflammation in different compartments, with a focus on the lung [80].

VLMS is a disease that affects children who (accidentally) ingest embryonated eggs of *T. canis* [80,84], a dog parasite [85]. Stage L3 larvae leave the egg in the intestine, travel to the liver through the circulatory system, breaching the venous sinusoids and hepatic parenchyma, or migrate directly from the hepatic vein to the inferior vena cava. Some larvae can cross the lymphatic vessels and reach the lung. From the lung, through the pulmonary vein, the larvae travel to the left atrium and enter the circulatory system. Although the mechanism is not entirely clear, it is known that the larvae, instead of completing their developmental cycle by returning to the intestine, can spread throughout the organism, where they encyst and are destroyed, or continue to migrate throughout tissues for many months, despite the humoral immune response [84]. As they migrate through tissues, the larvae release metabolites and surface antigens [86], initiating an inflammatory response that has as its principal characteristics intense eosinophilic infiltration of different tissues [65,87,88] and high IgE levels [89,90]. Such characteristics are also present in allergic diseases [91] and in asthma [92]. Although clinical reports suggest a relationship between asthma and toxocariasis, little is known about how parasitic infections predispose or protect individuals against the development of asthma [93]. Buijs and collaborators (1994) have demonstrated that *T. canis* infection induces eosinophilia in the airways, resulting in reduced pulmonary capacity [93]. These authors suggest that *T. canis* antigens act as polyclonal B lymphocyte activators, leading to increased IgE synthesis specific for the parasite and other unrelated antigens, predisposing the patient to asthma [93]. The authors detected IgE specific for *T. canis* in children with asthma and allergic diseases [94] therefore, in certain asthma cases, evaluation of levels of antibodies to *T. canis* or other parasites becomes indispensable [93]. On the other hand, other observations suggest that the increased IgE production in helminthic infections is capable of blocking the allergic response to unrelated antigens. In such a case, the IgE molecules specific for the parasite would occupy the FcRI receptors and would hinder the specific linking of specific IgE targeted to the allergens [89,95]. The blocking activity of parasite-specific IgE, in the context of allergic reactions, was demonstrated in animal models and described in patients [89,95].

Various studies have demonstrated that high IL-4 concentrations in the lungs are indicative of asthmatic symptoms, in addition to being chemotactic for eosinophils [96]. Thus, the recruited eosinophils play a fundamental role in the inflammatory reaction, inducing asthmatic symptoms, as described above. After being activated, these cells can degranulate, liberating granular constituents such as MBP that are toxic and cause scaling and bronchial epithelial destruction, contributing to airway hyperreactivity [97]. This pathway also liberates leukotrienes and PAF, which contribute to bronchial constriction, increasing vascular permeability and intensifying bronchial hyperreactivity, as well as activating more eosinophils [98].

We evaluated the effect of pcDNA-IL-12, a plasmid expressing the p35 and p40 subunits of murine IL-12 and pcDNA3-CPG (Invitrogen, Carlsbad, CA, USA), on eosinophilic inflammation, hyperreactivity of the air ways, antibody production, and cytokine production, in a murine model of *T. canis* infection. It was observed, for the first time, that the parasitic infection increased bronchial reactivity, although this mechanism is not well understood [80]. The data presented in this article demonstrated that pcDNA-IL-12 blocked eosinophilia in the blood and the bronchoalveolar lavage fluid (BALF), as well as pulmonary eosinophilic inflammation. Moreover, pcDNA-IL-12 shifted the Th2-dominant cytokine profile to one dominated by Th1, leading to the production of IgG1 antibodies and increased IL-12 production. We did not observe any impact of pcDNA-IL-12 on airway hyperreactivity, as has been described by other authors [49,56]. The findings of this work permitted us to infer that the administration route (gene gun) could interfere with the response induced by the plasmid and its effect on pulmonary hyperreactivity. According to data from the literature, the systemic administration of IL-12 was efficient in inhibiting eosinophilia and blocking pulmonary hyperreactivity [99]. According to Lee et al., (2001) [49], as reviewed by Malheiro et al., 2008 [80], it has been demonstrated that direct administration of the IL-12 gene within airways inhibits eosinophilia and bronchial hyperresponsivity. Thus, the authors suggested that the hyperreactivity observed in *T. canis* infection probably involves a more complex and heterogeneous mechanism.

Recent data in the literature demonstrate that the balance of regulatory T cells and standard Th2 cytokines is fundamental for the development of immunity to allergens. In asthma, dendritic cells stimulated with DNA vaccine also reduce the Th2 response as well as the expression of co-stimulatory molecules. Recently, the participation of regulatory T cells following vaccination with pcDNA-Fc-OVA was demonstrated in a murine model of respiratory allergy [100]. It is well established that IL-12 contributes to the immune response by inhibiting Th2 expression. We demonstrate that pcDNA-IL-12 inhibited the production of IL-4 and IL-10, but in a delayed fashion, 48 days after infection with *T. canis*. On the other hand, we demonstrated that pcDNA-3-CPG blocked the Th2 response, by 24 days after the infection. This effect was more prominent after 48 days, as was seen for pcDNA-IL-12. Curiously, both plasmids induced IL-12 production. We suggest that this effect is related to CPG domains present in these plasmids that are known to activate murine, human and bovine cells, thereby inducing the production of inflammatory cytokines that act on mononuclear and dendritic cells [101]. Moreover, airway hyperreactivity can occur in the absence of eosinophilia, as pcDNA-IL-12 inhibits eosinophilia, but did not affect the reduction in bronchial hyperreactivity, as previously reported. On the other hand, we also demonstrate that pcDNA-3 inhibits hyperreactivity, but that it does not completely eliminate the presence of eosinophils, thus confirming the previous hypothesis. The results presented in the article cited and reviewed here illustrate that other relevant elements may regulate the local immune response, with effects on airway hyperreactivity. It is important to mention that asthma-related vaccine research seeks to address an urgent problem that affects millions of people, but requires the knowledge of a complex immune mechanism, which can involve regulatory T cells and/or Th17, as well as other mediators. Therefore, further studies will be necessary to elucidate the pathophysiology of asthma in different experimental models.

## References


[1] Lowrie DB, Tascon RE, Bonato VL, Lima VM, Faccioli LH, Stavropoulos E et al. Therapy of tuberculosis in mice by DNA vaccination. *Nature* 1999; 400(6741):269-271.

[2] Dietrich G, Kolb-Maurer A, Spreng S, Schartl M, Goebel W, Gentschev I. Gram-positive and Gram-negative bacteria as carrier systems for DNA vaccines. *Vaccine* 2001; 19(17-19):2506-2512.

[3] Ulmer JB, Donnelly JJ, Parker SE, Rhodes GH, Felgner PL, Dwarki VJ et al. Heterologous protection against influenza by injection of DNA encoding a viral protein. *Science* 1993; 259(5102):1745-1749.

[4] Spier RE. International meeting on the nucleic acid vaccines for the prevention of infectious disease and regulating nucleic acid (DNA) vaccines. Natcher Conference Center NIH, Bethesda, MD 5-8 February, 1996. *Vaccine* 1996; 14(13):1285-1288.

[5] Corr M, Lee DJ, Carson DA, Tighe H. Gene vaccination with naked plasmid DNA: mechanism of CTL priming. *J. Exp. Med.* 1996; 184(4):1555-1560.

[6] Akbari O, Panjwani N, Garcia S, Tascon R, Lowrie D, Stockinger B. DNA vaccination: transfection and activation of dendritic cells as key events for immunity. *J. Exp. Med.* 1999; 189(1):169-178.

[7] Gurunathan S, Klinman DM, Seder RA. DNA vaccines: immunology, application, and optimization*. *Annu. Rev. Immunol.* 2000; 18:927-974.

[8] Krug A, Towarowski A, Britsch S, Rothenfusser S, Hornung V, Bals R et al. Toll-like receptor expression reveals CpG DNA as a unique microbial stimulus for plasmacytoid dendritic cells which synergizes with CD40 ligand to induce high amounts of IL-12. *Eur. J. Immunol.*2001; 31(10):3026-3037.

[9] Zelenay S, Elias F, Flo J. Immunostimulatory effects of plasmid DNA and synthetic oligodeoxynucleotides. *Eur. J. Immunol.* 2003; 33(5):1382-1392.

[10] Ishii KJ, Akira S. Innate immune recognition of, and regulation by, DNA. *Trends Immunol.* 2006; 27(11):525-532.

[11] Krieg AM, Yi AK, Matson S, Waldschmidt TJ, Bishop GA, Teasdale R et al. CpG motifs in bacterial DNA trigger direct B-cell activation. *Nature* 1995; 374(6522):546-549.

[12] Krieg AM. Now I know my CpGs. *Trends Microbiol.* 2001; 9(6):249-252.

[13] Krieg AM. Immune effects and mechanisms of action of CpG motifs. *Vaccine* 2000; 19(6):618-622.

[14] Kovarik J, Bozzotti P, Tougne C, Davis HL, Lambert PH, Krieg AM et al. Adjuvant effects of CpG oligodeoxynucleotides on responses against T-independent type 2 antigens. *Immunology* 2001; 102(1):67-76.
[15] Wild JS, Sur S. CpG oligonucleotide modulation of allergic inflammation. *Allergy* 2001; 56(5):365-376.
[16] Donnelly JJ, Ulmer JB, Shiver JW, Liu MA. DNA vaccines. *Annu. Rev. Immunol.* 1997; 15:617-648.
[17] Franco LH, Wowk PF, Silva CL, Trombone AP, Coelho-Castelo AA, Oliver C et al. A DNA vaccine against tuberculosis based on the 65 kDa heat-shock protein differentially activates human macrophages and dendritic cells. *Genet. Vaccines Ther.* 2008; 6:3.
[18] Nchinda G, Kuroiwa J, Oks M, Trumpfheller C, Park CG, Huang Y et al. The efficacy of DNA vaccination is enhanced in mice by targeting the encoded protein to dendritic cells. *J. Clin. Invest.* 2008; 118(4):1427-1436.
[19] Chono S, Tanino T, Seki T, Morimoto K. Influence of particle size on drug delivery to rat alveolar macrophages following pulmonary administration of ciprofloxacin incorporated into liposomes. *J. Drug Target* 2006; 14(8):557-566.
[20] Haynes JR. Genetic vaccines. *Infect. Dis. Clin. North Am.* 1999; 13(1):11-26, v.
[21] Levy MY, Barron LG, Meyer KB, Szoka FC, Jr. Characterization of plasmid DNA transfer into mouse skeletal muscle: evaluation of uptake mechanism, expression and secretion of gene products into blood. *Gene Ther.* 1996; 3(3):201-211.
[22] Lai WC, Bennett M, Johnston SA, Barry MA, Pakes SP. Protection against Mycoplasma pulmonis infection by genetic vaccination. *DNA Cell Biol.* 1995; 14(7):643-651.
[23] Leitner WW, Seguin MC, Ballou WR, Seitz JP, Schultz AM, Sheehy MJ et al. Immune responses induced by intramuscular or gene gun injection of protective deoxyribonucleic acid vaccines that express the circumsporozoite protein from Plasmodium berghei malaria parasites. *J. Immunol.* 1997; 159(12):6112-6119.
[24] Vanderzanden L, Bray M, Fuller D, Roberts T, Custer D, Spik K et al. DNA vaccines expressing either the GP or NP genes of Ebola virus protect mice from lethal challenge. *Virology* 1998; 246(1):134-144.
[25] Pertmer TM, Eisenbraun MD, McCabe D, Prayaga SK, Fuller DH, Haynes JR. Gene gun-based nucleic acid immunization: elicitation of humoral and cytotoxic T lymphocyte responses following epidermal delivery of nanogram quantities of DNA. *Vaccine* 1995; 13(15):1427-1430.
[26] Feltquate DM, Heaney S, Webster RG, Robinson HL. Different T helper cell types and antibody isotypes generated by saline and gene gun DNA immunization. *J. Immunol.* 1997; 158(5):2278-2284.
[27] Jilek S, Walter E, Merkle HP, Corthesy B. Modulation of allergic responses in mice by using biodegradable poly(lactide-co-glycolide) microspheres. *J. Allergy Clin. Immunol.* 2004; 114(4):943-950.
[28] de Paula L, Silva CL, Carlos D, Matias-Peres C, Sorgi CA, Soares EG et al. Comparison of different delivery systems of DNA vaccination for the induction of protection against tuberculosis in mice and guinea pigs. *Genet. Vaccines Ther.* 2007; 5:2.
[29] Memarnejadian A, Roohvand F. Fusion of HBsAg and prime/boosting augment Th1 and CTL responses to HCV polytope DNA vaccine. *Cell Immunol.* 2010; 261(2):93-98.

[30] Jiang G, Shi M, Conteh S, Richie N, Banania G, Geneshan H et al. Sterile protection against Plasmodium knowlesi in rhesus monkeys from a malaria vaccine: comparison of heterologous prime boost strategies. *PLoS. One* 2009; 4(8):e6559.

[31] Wonglertnirant N, Todo H, Opanasopit P, Ngawhirunpat T, Sugibayashi K. Macromolecular delivery into skin using a hollow microneedle. *Biol. Pharm. Bull.* 2010; 33(12):1988-1993.

[32] Rosada RS, de la Torre LG, Frantz FG, Trombone AP, Zarate-Blades CR, Fonseca DM et al. Protection against tuberculosis by a single intranasal administration of DNA-hsp65 vaccine complexed with cationic liposomes. *BMC Immunol.* 2008; 9:38.

[33] Muller RH, Mader K, Gohla S. Solid lipid nanoparticles (SLN) for controlled drug delivery -a review of the state of the art. *Eur. J. Pharm. Biopharm.* 2000; 50(1):161-177.

[34] Wolf SF, Temple PA, Kobayashi M, Young D, Dicig M, Lowe L et al. Cloning of cDNA for natural killer cell stimulatory factor, a heterodimeric cytokine with multiple biologic effects on T and natural killer cells. *J. Immunol.* 1991; 146(9):3074-3081.

[35] Gubler U, Chua AO, Schoenhaut DS, Dwyer CM, McComas W, Motyka R et al. Coexpression of two distinct genes is required to generate secreted bioactive cytotoxic lymphocyte maturation factor. *Proc. Natl. Acad. Sci. U S A* 1991; 88(10):4143-4147.

[36] Gately MK, Wolitzky AG, Quinn PM, Chizzonite R. Regulation of human cytolytic lymphocyte responses by interleukin-12. *Cell Immunol.* 1992; 143(1):127-142.

[37] Wills-Karp M. IL-12/IL-13 axis in allergic asthma. *J. Allergy Clin. Immunol.* 2001; 107(1):9-18.

[38] D'Andrea A, Rengaraju M, Valiante NM, Chehimi J, Kubin M, Aste M et al. Production of natural killer cell stimulatory factor (interleukin 12) by peripheral blood mononuclear cells. *J. Exp. Med.* 1992; 176(5):1387-1398.

[39] Kobayashi M, Fitz L, Ryan M, Hewick RM, Clark SC, Chan S et al. Identification and purification of natural killer cell stimulatory factor (NKSF), a cytokine with multiple biologic effects on human lymphocytes. *J. Exp. Med.* 1989; 170(3):827-845.

[40] Schoenhaut DS, Chua AO, Wolitzky AG, Quinn PM, Dwyer CM, McComas W et al. Cloning and expression of murine IL-12. *J. Immunol.* 1992; 148(11):3433-3440.

[41] Chan SH, Perussia B, Gupta JW, Kobayashi M, Pospisil M, Young HA et al. Induction of interferon gamma production by natural killer cell stimulatory factor: characterization of the responder cells and synergy with other inducers. *J. Exp. Med.* 1991; 173(4):869-879.

[42] Perussia B, Chan SH, D'Andrea A, Tsuji K, Santoli D, Pospisil M et al. Natural killer (NK) cell stimulatory factor or IL-12 has differential effects on the proliferation of TCR-alpha beta+, TCR-gamma delta+ T lymphocytes, and NK cells. *J. Immunol.* 1992; 149(11):3495-3502.

[43] Trinchieri G. Interleukin-12 and its role in the generation of TH1 cells. *Immunol. Today* 1993; 14(7):335-338.

[44] Naume B, Johnsen AC, Espevik T, Sundan A. Gene expression and secretion of cytokines and cytokine receptors from highly purified CD56+ natural killer cells stimulated with interleukin-2, interleukin-7 and interleukin-12. *Eur. J. Immunol.* 1993; 23(8):1831-1838.

[45] Manetti R, Parronchi P, Giudizi MG, Piccinni MP, Maggi E, Trinchieri G et al. Natural killer cell stimulatory factor (interleukin 12 [IL-12]) induces T helper type 1 (Th1)-

specific immune responses and inhibits the development of IL-4-producing Th cells. *J. Exp. Med.* 1993; 177(4):1199-1204.

[46] Aste-Amezaga M, D'Andrea A, Kubin M, Trinchieri G. Cooperation of natural killer cell stimulatory factor/interleukin-12 with other stimuli in the induction of cytokines and cytotoxic cell-associated molecules in human T and NK cells. *Cell Immunol.* 1994; 156(2):480-492.

[47] McKnight AJ, Zimmer GJ, Fogelman I, Wolf SF, Abbas AK. Effects of IL-12 on helper T cell-dependent immune responses in vivo. *J. Immunol.* 1994; 152(5):2172-2179.

[48] Marshall JD, Secrist H, DeKruyff RH, Wolf SF, Umetsu DT. IL-12 inhibits the production of IL-4 and IL-10 in allergen-specific human CD4+ T lymphocytes. *J. Immunol.* 1995; 155(1):111-117.

[49] Lee SY, Huang CK, Zhang TF, Schofield BH, Burks AW, Bannon GA et al. Oral administration of IL-12 suppresses anaphylactic reactions in a murine model of peanut hypersensitivity. *Clin. Immunol.* 2001; 101(2):220-228.

[50] Marinaro M, Boyaka PN, Finkelman FD, Kiyono H, Jackson RJ, Jirillo E et al. Oral but not parenteral interleukin (IL)-12 redirects T helper 2 (Th2)-type responses to an oral vaccine without altering mucosal IgA responses. *J. Exp. Med.* 1997; 185(3):415-427.

[51] Gavett SH, O'Hearn DJ, Li X, Huang SK, Finkelman FD, Wills-Karp M. Interleukin 12 inhibits antigen-induced airway hyperresponsiveness, inflammation, and Th2 cytokine expression in mice. *J. Exp. Med.* 1995; 182(5):1527-1536.

[52] Kim TS, DeKruyff RH, Rupper R, Maecker HT, Levy S, Umetsu DT. An ovalbumin-IL-12 fusion protein is more effective than ovalbumin plus free recombinant IL-12 in inducing a T helper cell type 1-dominated immune response and inhibiting antigen-specific IgE production. *J. Immunol.* 1997; 158(9):4137-4144.

[53] Kips JC, Brusselle GJ, Joos GF, Peleman RA, Tavernier JH, Devos RR et al. Interleukin-12 inhibits antigen-induced airway hyperresponsiveness in mice. *Am. J. Respir. Crit. Care Med.* 1996; 153(2):535-539.

[54] Wynn TA, Eltoum I, Oswald IP, Cheever AW, Sher A. Endogenous interleukin 12 (IL-12) regulates granuloma formation induced by eggs of Schistosoma mansoni and exogenous IL-12 both inhibits and prophylactically immunizes against egg pathology. *J. Exp. Med* .1994; 179(5):1551-1561.

[55] Finkelman FD, Boyce JA, Vercelli D, Rothenberg ME. Key advances in mechanisms of asthma, allergy, and immunology in 2009. *J. Allergy Clin. Immunol.* 2010; 125(2):312-318.

[56] Stampfli MR, Scott NG, Wiley RE, Cwiartka M, Ritz SA, Hitt MM et al. Regulation of allergic mucosal sensitization by interleukin-12 gene transfer to the airway. *Am. J. Respir. Cell Mol. Biol.* 1999; 21(3):317-326.

[57] Gazzinelli RT, Giese NA, Morse HC, III. In vivo treatment with interleukin 12 protects mice from immune abnormalities observed during murine acquired immunodeficiency syndrome (MAIDS). *J. Exp. Med.* 1994; 180(6):2199-2208.

[58] Melero I, Mazzolini G, Narvaiza I, Qian C, Chen L, Prieto J. IL-12 gene therapy for cancer: in synergy with other immunotherapies. *Trends Immunol* .2001; 22(3):113-115.

[59] Scott P, Trinchieri G. IL-12 as an adjuvant for cell-mediated immunity. *Semin. Immunol.* 1997; 9(5):285-291.

[60] Afonso LC, Scharton TM, Vieira LQ, Wysocka M, Trinchieri G, Scott P. The adjuvant effect of interleukin-12 in a vaccine against Leishmania major. *Science* 1994; 263(5144):235-237.

[61] Gazzinelli RT, Hieny S, Wynn TA, Wolf S, Sher A. Interleukin 12 is required for the T-lymphocyte-independent induction of interferon gamma by an intracellular parasite and induces resistance in T-cell-deficient hosts. *Proc. Natl. Acad. Sci. U S A* 1993; 90(13):6115-6119.

[62] Boyce JA, Friend D, Matsumoto R, Austen KF, Owen WF. Differentiation in vitro of hybrid eosinophil/basophil granulocytes: autocrine function of an eosinophil developmental intermediate. *J. Exp. Med.* 1995; 182(1):49-57.

[63] Sanderson CJ, Warren DJ, Strath M. Identification of a lymphokine that stimulates eosinophil differentiation in vitro. Its relationship to interleukin 3, and functional properties of eosinophils produced in cultures. *J. Exp. Med.* 1985; 162(1):60-74.

[64] Metcalf D. The molecular biology and functions of the granulocyte-macrophage colony-stimulating factors. *Blood* 1986; 67(2):257-267.

[65] Faccioli LH, Mokwa VF, Silva CL, Rocha GM, Araujo JI, Nahori MA et al. IL-5 drives eosinophils from bone marrow to blood and tissues in a guinea-pig model of visceral larva migrans syndrome. *Mediators Inflamm.* 1996; 5(1):24-31.

[66] Dent LA, Strath M, Mellor AL, Sanderson CJ. Eosinophilia in transgenic mice expressing interleukin 5. *J. Exp. Med.* 1990; 172(5):1425-1431.

[67] Tominaga A, Takaki S, Hitoshi Y, Takatsu K. Role of the interleukin 5 receptor system in hematopoiesis: molecular basis for overlapping function of cytokines. *Bioessays* 1992; 14(8):527-533.

[68] Lewis CC, Aronow B, Hutton J, Santeliz J, Dienger K, Herman N et al. Unique and overlapping gene expression patterns driven by IL-4 and IL-13 in the mouse lung. *J. Allergy Clin. Immunol.* 2009; 123(4):795-804.

[69] WHO, Global Intiative for Asthma Global strategy for asthma management and prevention NBLBJ/WHO workshop report. 1995. Ref Type: Generic

[70] Mould AW, Matthaei KI, Young IG, Foster PS. Relationship between interleukin-5 and eotaxin in regulating blood and tissue eosinophilia in mice. *J. Clin. Invest.* 1997; 99(5):1064-1071.

[71] Hogan SP, Rosenberg HF, Moqbel R, Phipps S, Foster PS, Lacy P et al. Eosinophils: biological properties and role in health and disease. *Clin. Exp. Allergy* 2008; 38(5):709-750.

[72] Minai-Fleminger Y, Levi-Schaffer F. Mast cells and eosinophils: the two key effector cells in allergic inflammation. *Inflamm. Res.* 2009; 58(10):631-638.

[73] Azzawi M, Bradley B, Jeffery PK, Frew AJ, Wardlaw AJ, Knowles G et al. Identification of activated T lymphocytes and eosinophils in bronchial biopsies in stable atopic asthma. *Am. Rev. Respir. Dis.* 1990; 142(6 Pt 1):1407-1413.

[74] Bousquet J, Chanez P, Lacoste JY, Barneon G, Ghavanian N, Enander I et al. Eosinophilic inflammation in asthma. *N. Engl. J. Med.* 1990; 323(15):1033-1039.

[75] Yamaguchi Y, Matsui T, Kasahara T, Etoh S, Tominaga A, Takatsu K et al. In vivo changes of hemopoietic progenitors and the expression of the interleukin 5 gene in eosinophilic mice infected with Toxocara canis. *Exp. Hematol.* 1990; 18(11):1152-1157.

[76] Steel C, Nutman TB. Regulation of IL-5 in onchocerciasis. A critical role for IL-2. *J. Immunol.* 1993; 150(12):5511-5518.

[77] Mosmann TR, Sad S. The expanding universe of T-cell subsets: Th1, Th2 and more. *Immunol. Today* 1996; 17(3):138-146.

[78] Anibal FF, Rogerio AP, Malheiro A, Machado ER, Martins-Filho OA, Andrade MC et al. Impact of MK886 on eosinophil counts and phenotypic features in toxocariasis. *Scand. J. Immunol.* 2007; 65(4):344-352.

[79] Kuribayashi K, Kodama T, Okamura H, Sugita M, Matsuyama T. Effects of post-inhalation treatment with interleukin-12 on airway hyper-reactivity, eosinophilia and interleukin-18 receptor expression in a mouse model of asthma. *Clin. Exp. Allergy* 2002; 32(4):641-649.

[80] Malheiro A, Anibal FF, Martins-Filho OA, Teixeira-Carvalho A, Perini A, Martins MA et al. pcDNA-IL-12 vaccination blocks eosinophilic inflammation but not airway hyperresponsiveness following murine Toxocara canis infection. *Vaccine* 2008; 26(3):305-315.

[81] Walsh GM. Targeting airway inflammation: novel therapies for the treatment of asthma. *Curr. Med. Chem.* 2006; 13(25):3105-3111.

[82] Bruselle GG, Kips JC, Peleman RA, Joos GF, Devos RR, Tavernier JH et al. Role of IFN-gamma in the inhibition of the allergic airway inflammation caused by IL-12. Am *J. Respir. Cell Mol. Biol.* 1997; 17(6):767-771.

[83] Sano K, Haneda K, Tamura G, Shirato K. Ovalbumin (OVA) and Mycobacterium tuberculosis bacilli cooperatively polarize anti-OVA T-helper (Th) cells toward a Th1-dominant phenotype and ameliorate murine tracheal eosinophilia. *Am. J. Respir. Cell Mol. Biol.* 1999; 20(6):1260-1267.

[84] Gomes de Moraes. *Parasitologia Médica,* first edition. 1971. Ref Type: Generic

[85] BEAVER PC, SNYDER CH, CARRERA GM, DENT JH, LAFFERTY JW. Chronic eosinophilia due to visceral larva migrans; report of three cases. *Pediatrics* 1952; 9(1):7-19.

[86] Rockey JH, John T, Donnelly JJ, McKenzie DF, Stromberg BE, Soulsby EJ. In vitro interaction of eosinophils from ascarid-infected eyes with Ascaris suum and Toxocara canis larvae. *Invest. Ophthalmol. Vis. Sci.* 1983; 24(10):1346-1357.

[87] Sugane K, Oshima T. Recovery of large numbers of eosinophils from mice infected with Toxocara canis. *Am. J. Trop. Med. Hyg.* 1980; 29(5):799-802.

[88] Kayes SG, Oaks JA. Toxocara canis: T lymphocyte function in murine visceral larva migrans and eosinophilia onset. *Exp. Parasitol.* 1980; 49(1):47-55.

[89] Turner KJ, Fisher EH, Mayrhofer G. Age-dependent modulation of serum IgE and mast cell sensitization by Nippostrongylus brasiliensis infestation in rats. *Aust. J. Exp. Biol. Med. Sci.* 1981; 59(4):491-502.

[90] Nutman TB, Hussain R, Ottesen EA. IgE production in vitro by peripheral blood mononuclear cells of patients with parasitic helminth infections. *Clin. Exp. Immunol.* 1984; 58(1):174-182.

[91] Kay AB, Moqbel R, Durham SR, MacDonald AJ, Walsh GM, Shaw RJ et al. Leucocyte activation initiated by IgE-dependent mechanisms in relation to helminthic parasitic disease and clinical models of asthma. *Int. Arch. Allergy Appl. Immunol.* 1985; 77(1-2):69-72.

[92] de Monchy JG, Keyzer JJ, Kauffman HF, Beaumont F, de Vries K. Histamine in late asthmatic reactions following house-dust mite inhalation. *Agents Actions* 1985; 16(3-4):252-255.

[93] Buijs J, van Knapen F. Toxocara infection in children and the relation with allergic manifestations. *Vet. Q 1994; 16 Suppl.* 1:13S-14S.

[94] Oteifa NM, Moustafa MA, Elgozamy BM. Toxocariasis as a possible cause of allergic diseases in children. *J. Egypt Soc. Parasitol.* 1998; 28(2):365-372.

[95] Weiss ST. Parasites and asthma/allergy: what is the relationship? *J. Allergy Clin. Immunol.* 2000; 105(2 Pt 1):205-210.

[96] Viola JP, Kiani A, Bozza PT, Rao A. Regulation of allergic inflammation and eosinophil recruitment in mice lacking the transcription factor NFAT1: role of interleukin-4 (IL-4) and IL-5. *Blood* 1998; 91(7):2223-2230.

[97] Wardlaw AJ, Hay H, Cromwell O, Collins JV, Kay AB. Leukotrienes, LTC4 and LTB4, in bronchoalveolar lavage in bronchial asthma and other respiratory diseases. *J. Allergy Clin. Immunol.* 1989; 84(1):19-26.

[98] Niimi A, Amitani R, Suzuki K, Tanaka E, Murayama T, Kuze F. Eosinophilic inflammation in cough variant asthma. *Eur. Respir. J.* 1998; 11(5):1064-1069.

[99] Bryan SA, Leckie MJ, Hansel TT, Barnes PJ. Novel therapy for asthma. *Expert Opin. Investig. Drugs* 2000; 9(1):25-42.

[100] Wang Y, Bai C, Wang G, Wang D, Cheng X, Huang J et al. Protection against the allergic airway inflammation depends on the modulation of spleen dendritic cell function and induction of regulatory T cells in mice. *Genet. Vaccines Ther.* 2010; 8:2.

[101] Park ST, Kim KE, Na K, Kim Y, Kim TY. Effect of dendritic cells treated with CpG ODN on atopic dermatitis of Nc/Nga mice. *J. Biochem. Mol. Biol.* 2007; 40(4):486-493.

In: DNA Vaccines: Types, Advantages and Limitations
Editors: E. C. Donnelly and A. M. Dixon
ISBN 978-1-61324-444-9


*Chapter VII*

# Cured By DNA-Genetic Immunization in the Therapeutic Sector

***Fiona J. Baird and Peter M. Smooker***[3]
School of Applied Sciences, RMIT University, Bundoora 3083 Australia

## Abstract

Historically, immunization has been designed to prevent the onset of infectious disease, with vaccines acting as prophylactic agents. However, over the course of time immunization has evolved to include a therapeutic objective whereby individuals who have already contracted an infectious or neoplastic disease, and where traditional treatment options are limited, are vaccinated. With diseases such as HIV and tuberculosis the limited success of traditional chemotherapeutic and prophylactic approaches has resulted in the experimental advancement of therapeutic genetic immunization, where the host immune system is to be modulated to improve disease prognosis or ultimately, eradicate the infection. Furthermore, for cancer, a major non-infectious disease, similar experimental treatments are underway. Here we will explore how genetic immunization enables re-direction of the host immune system resulting in the development of effective immune responses.

**Keywords:** DNA vaccine, therapy, cancer, HIV, hepatitis, tuberculosis

## Introduction

Globally, vaccination has been successful against many infectious diseases throughout the last fifty years. The eradication of smallpox using vaccinia immunization demonstrated that prophylactic vaccination could save lives and eliminate diseases on a global scale. The elucidation of disease pathogenesis has changed the approach for both treatment and

[3] Corresponding author: peter.smooker@rmit.edu.au.

vaccination. Bacterial and viral pathogens are easier targets for vaccination than autoimmune diseases and carcinoma, which generally have no foreign pathogens that instigate the onset of disease. The link between foreign pathogens and autoimmune diseases and carcinomas has only recently been defined for several diseases such as cervical cancer and papilloma virus (as well as other infections) [1], and Guillian-Barré syndrome and *Campylobacter jejuni* [2]. Recently the objective of vaccination has diverged into two different streams. The first is the traditional route of prophylaxis and the second is that of therapeutics (Table 1).

**Table 1. Comparison of prophylaxis and therapeutic genetic vaccinations Compiled from available regimen data on ClinicalTrials.gov**

| | Prophylaxis | Therapeutic |
|---|---|---|
| Doses administered | One dose (preferred) | Multiple doses |
| Dosage | < 1 mg per dose | > 1 mg per dose |
| Frequency of vaccination | Once in a lifetime (preferred) | Monthly |
| Administration | Needle, electroporation, etc. | Needle, electroporation, etc. |

Of the four currently licensed veterinary DNA vaccines two are for therapeutic applications. The first is ONCEPT™ by Merial, used to treat melanoma in dogs. ONCEPT™ contains the gene for human tyrosinase which is an enzyme associated with skin pigmentation and is commonly made by melanoma tumors. This therapeutic vaccine is given as a treatment once the dog has been diagnosed with malignant melanoma and has shown great efficacy as the test subjects lived significantly longer with the treatment than without (Merial press release, February 16, 2010). Preclinical trials showed that the treatment greatly improved the quality of life of the subjects even though they were in metastatic stages of the disease [3]. This is a traditional therapeutic where the application is to diminish symptoms and improve the disease prognosis.

The second licensed therapeutic DNA vaccine is LifeTide® SW 5 by VGX Animal Health that is used once in sows of breeding age to increase the number of piglets. The DNA vaccine encodes porcine GNRH, and according to a study conducted by VGX, when sows are treated with LifeTide® SW 5 their piglets were heavier in birth weight and at weaning, and the meat was of better quality compared to the controls [4]. This therapeutic genetic application is to improve the breeding outcome of a food product so that more progeny can be obtained from fewer sows without increasing maintenance or production costs. Clearly, this is an economic imperative and is different than the aim of vaccines developed for the medical sector.

In the veterinary industry, therapeutic DNA vaccines will continue to have dual purposes of disease treatment and product improvement. However therapeutic DNA vaccines for medical use in humans are focusing on treatment of various diseases via genetic immunization. Despite there being no licensed DNA vaccines for human use, neither prophylactic nor therapeutic, there are multiple vaccines in trials to establish efficacy and safety. These vaccines are aimed at treating chronic conditions such as herpes, tuberculosis (TB), hepatitis, HIV and different forms of cancer. These diseases have limited treatment options and traditional vaccines have not been successfully manufactured. The goal of immunotherapy using genetic immunization in chronic illnesses is to restore immune-competence and reverse the anergy-like state of the patient's immune system (Figure 1) [5].

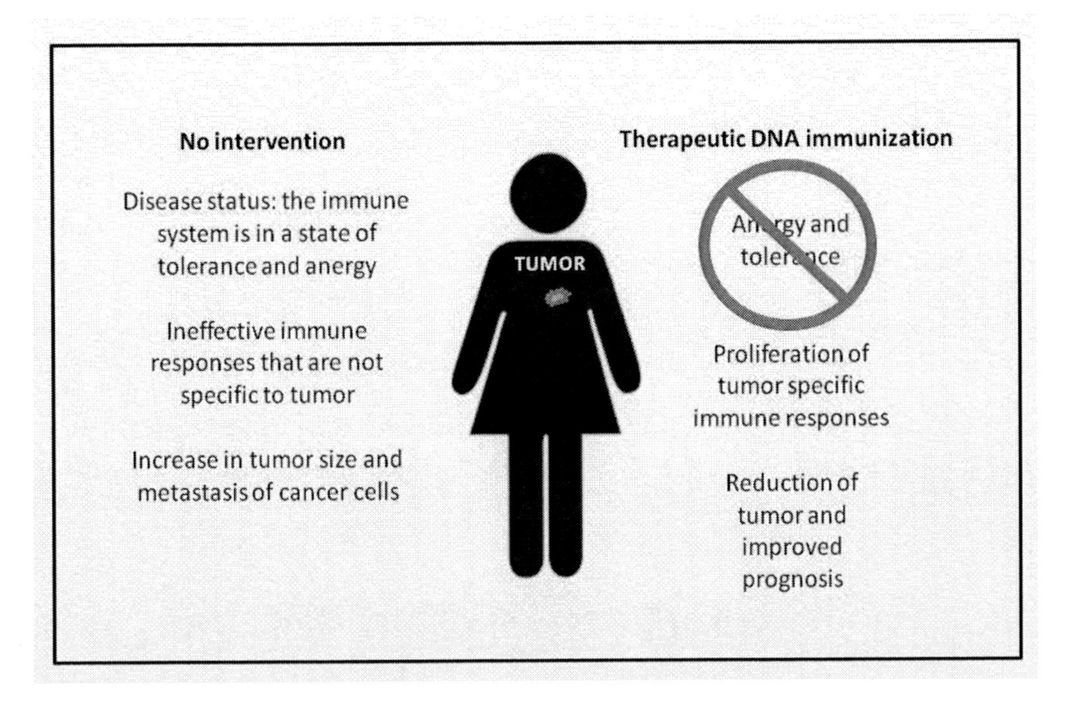


Figure 1. Implications of genetic immunization intervention in the example of cancer.

## Modulation of the Immune System

Genetic vaccination can elicit both cellular and humoral immune responses however these responses need to be modulated depending on the disease. For example, stimulating the humoral immune responses in chronic TB cases can exacerbate the severity of the disease [6]. The continuing investigation into how to use genetic immunization to target specific immune cells is being carried out in both animal trials and human clinical trials for many human diseases.

The uses of adoptive T cell therapy in cancer treatment where autologous antigen-specific T cells are proliferated in tissue culture and re-infused to patients to activate the patient's immune system against the cancer has been successful over the past decade. However, these therapies are expensive, and are difficult to maintain and administer on a large scale. Using genetic vaccination to induce clonal expansion of T cells *in vivo* is an ongoing aim of researchers. The aim is to facilitate the induction of mature antigen-presenting cells (APCs) and dendritic cells (DCs) to activate more T cells with further differentiation into both effector and helper subsets, whilst overcoming the network of regulatory T (Treg) cells that are often induced by tumors [7]. The DNA vaccines must elicit specific immune subsets into action which will lead to a potent and appropriate response. For example, the failure of the STEP trial [8] where the group who received an adenovirus-vectored DNA vaccine and had high pre-existing immunity to the vector showed significantly higher rates of new HIV infections than the placebo group, which could be attributed to a subset of memory CD4+ T cells being elicited by the adenovirus that was more susceptible to HIV [9-10]. Another example is hepatitis where suppression of the innate immune responses in mice by hepatitis B virus (HBV) has recently been reported [11], so therapeutic immunizations should not only be targeted at the hepatocytes but to activate the innate immune responses to ensure a comprehensive immune response is mounted against the infection.

The interaction between immune cells and DNA vaccines is yet to be fully elucidated. Myocytes are the main cell type transfected during DNA vaccination [12] and cross-priming from myocytes to APCs is the main pathway for CTL proliferation [13]. Only a small percentage of APCs are directly transfected during vaccination [12]. This ability to bypass the traditional MHCI restricted pathway for proliferation of immune responses can be useful in such cases where the disease causes anergy, for example, in cancers. One of the striking benefits of DNA vaccines is their ability to stimulate not only the CTLs but the T-helper and humoral branches of the immune system, as this can result in a broad, longlasting and effective immune response. However the disease pathology must be fully elucidated, and taken into consideration when designing vaccines so that eliciting immune responses that could potentially favor the disease can be avoided.

# The Big Four – Cancer, HIV, Hepatitis and Tuberculosis

## Cancer

Globally, cancer, in all its manifestations, is a leading cause of death and is predicted to carry on increasing to 12 million deaths in 2030 [14]. According to the World Health Organization (WHO), 30% of all cancers are preventable with lifestyle changes and early detection. According to the U.S. National Cancer Institute, breast and prostate cancers receive the most research funding [15] even though lung and stomach cancers are more prevalent [14]. Traditionally chemotherapy, radiotherapy along with surgical intervention has been the treatment of choice for most cancers; however the development of DNA vaccines into therapeutics has lead to new experimental treatments.

The most publicized DNA vaccines to reach clinical trials are for prostate cancer. Currently, there are two phase II trials for therapeutic DNA agents against prostate cancer – one encoding Prostatic Acid Phosphatase (PAP) (ClinicalTrials.gov ID: NCT00849121) and one encoding rhesus prostate specific antigen (PSA) (ClinicalTrials.gov ID: NCT00859729). These two trials are due for completion in 2014 and 2011, respectively; however researchers are still evaluating different antigens and modes of administration to increase the immunogenicity of these vaccines. Plasmid expressing human PSA has shown great promise in mice where treatment delayed the emergence of tumors and resulted in prolonged survival. These responses were stimulated from four doses delivered by electroporation and resulted in tumor-specific immune responses which were successfully used in adoptive T cell transfer experiments [16].

Other clinical studies include a phase I study evaluating a mammaglobin-A DNA vaccine given as a naked plasmid DNA vaccine to women with metastatic breast cancer. The study's objective is to demonstrate that the mammaglobin-A DNA vaccine can be safely administered in humans and generate measurable CD8 T cell responses to mammaglobin-A (ClinicalTrials.gov Identifier: NCT00807781). Another phase I clinical trial is using a therapeutic vaccine consisting of recombinant DNA and adenovirus expressing L523S protein in patients with early stage non-small cell lung cancer (ClinicalTrials.gov ID: NCT00062907). This trial will determine if this vaccine is efficacious and safe for administration in patients with lung cancer. A recent phase I trial to commence is the use of a DNA vaccine expressing a modified antibody delivered by electroporation which should activate proliferation of T cells that are melanoma-specific which will lead to the elimination of melanoma cells (ClinicalTrials.gov ID: NCT01138410).

In terms of mode of action, to prevent tumor development a DNA vaccine may target angiogenesis which directly enables tumor growth. Multiple studies have reported various levels of success in targeting this pathway. A novel DNA vaccine which delivered listeriolysin O-fetal liver kinase to tumors slowed tumor growth and the tumors displayed reduced vasculature in mice [17]. Another study has targeted the delta-like ligand 4 (DLL4) which is an important ligand expressed by endothelial tip cells during angiogenic sprouting and thought to be highly expressed during malignant growth. This DNA vaccine is aimed at inhibiting expression of DLL4 in patients with breast cancer where it is over-expressed in cancerous cells and preliminary trials in mice have demonstrated severely reduced growth of orthotopically implanted mammary carcinomas [18].

The use of different vector systems has also been applied to cancer therapeutics. One recent study conducted in humans used alphavirus packaged in virus-like replicon particles (VRP) expressing the tumor antigen carcinoembryonic antigen (CEA). These VRP were efficiently taken up by DCs, and were tolerated well by patients with metastatic cancer after multiple doses. Most promisingly they also overcame both high titers of neutralizing antibodies to alphavirus and elevated Treg levels to induce CEA-specific T cell and humoral responses [19]. An adeno-associated virus serotype 2 has been used as a vector in mice to deliver GM-CSF and B7-1 to growing tumors which resulted in significant reduction in tumor growth and prolonged survival. Stunningly, there were cured mice with induced specific anti-tumor T cell responses that were resistant to re-challenge [20].

One of the difficulties of evaluating these genetic therapies is the assessment criteria to which therapeutic vaccines are tested against. For conditions such as cancers the normal measures of immune responses such as antibody titers and cytokine responses should not be the sole indication of whether a DNA vaccine is improving or worsening a condition. When assessing trials run over the last decade the majority (>96%) of patients who received therapeutic vaccines did not exhibit evidence of cancer regression using standard oncologic reporting criteria [5]. In a similar vein, a new therapy, an APC-based vaccine called Sipuleucel-T has been approved by the Food and Drug Administration as the first therapeutic vaccine in humans even though it did not show short-term changes to the disease progression. However, the vaccine significantly improved overall survival by 4.8 months [21]. The criteria on which a therapeutic vaccine, including DNA therapies, is evaluated needs to be revised to include long term survival rates.

The evolution of therapeutic DNA immunization for cancer patients is not progressing as fast as researchers predicted, probably due to the self-nature of the disease. With further elucidation of specific tumor antigens and revision of the existing criteria on which DNA vaccines are evaluated against, more treatments will become available that improve prognosis and quality of life for cancer patients.

## HIV

Worldwide in 2009 there were over 33 million people living with HIV, 2.6 million people who contracted HIV and 1.8 million deaths as a result of AIDS [22]. The pathophysiology of HIV and its progression to AIDS is complex and traditional methods of treatment have failed to cure the infection. HIV has the ability to be latent in tissues for periods of time, and standard therapeutics cannot access the virus at these times. In addition to this, HIV is highly mutagenic and genetically exhibits multiple clades spread around the globe. The history of HIV and its treatments is reviewed in [23].

The use of highly active antiretroviral therapy (HAART) is the most common treatment for HIV patients particularly in early infection and as a cure has not been found by either prophylactic vaccination or the therapeutic route alone, combination of HAART and vaccination are currently being evaluated. One of the earliest studies to suggest combination therapy with both vaccination and pharmaceuticals was from Sandstrom and colleagues [24] who found that vaccination with recombinant gp160 alone had modest effects at eliciting CD4+ cells, however in combination with antiretroviral therapy elicited slightly higher levels. This combination treatment prolonged the period of stable CD4+ T cell levels in subjects before development of AIDS and death. These observations lead to DNA vaccination being explored as a combination therapy with traditional antiviral treatments.

Preclinical trials using simian immunodeficiency virus (SIV) as a model disease have shown that DNA vaccination can stimulate the immune system and reduce clinical symptoms. One study has used a mixture of plasmids to vaccinate macaques during antiretroviral therapy. This resulted in specific cellular and humoral immune responses to the Gag, Env, Nef, and Pol antigens as well as an approximate 1 log reduction in viremia [25]. This study indicates that the use of genetic immunization in tandem with traditional pharmaceutical therapies could result in more efficacious management of this chronic disease. This strategy was tested in a phase I/II trial using a DNA vaccine VRC-HIVDNA009-00-VP which encodes a HIV-1 subtype B Gag-Pol-Nef fusion protein and modified Env constructs from subtypes A, B and C in a 4 plasmid mixture. The participants of this trial underwent antiretroviral therapy in the acute/early phase of HIV infection along with the vaccination regime. Unfortunately there were no significant differences between the vaccinated groups and the placebo in terms of viremia but the vaccinations were found to be safe [26].

Standalone therapeutic DNA vaccines are still being investigated particularly for targeted applications. A DNA vaccine targeted to stimulate DC proliferation *in vivo* was evaluated using a canarypox vector encoding HIV-1 Env and Gag and a synthetic polypeptide encompassing epitopes from Nef and Pol delivered in DCs to the patient. Although the vaccine was demonstrated to be safe, it was not strongly immunogenic and efficacy did not differ from the placebo treatments [27]. These failures to demonstrate efficacy along with the well-documented failure of the Merck STEP trial [8] has not impeded further clinical trials. The PEDVAC trial is the first phase II trial of a therapeutic DNA vaccine being tested in HIV positive children. The vaccine is a multiclade, multigene HIV-DNA vaccine (HIVIS) designed to raise broad immune responses against several subtypes of the virus and preliminary safety data has shown that the vaccine is well-tolerated and halted the decrease of CD4+ T cell counts from baseline [28], which would have indicated treatment failure [29].

Just like cancer, HIV is a difficult disease to target for both prophylaxis and therapeutic vaccination purposes. However as it is a viral infectious disease the development of therapeutic DNA vaccines based on the conserved antigens between viral clades are still being evaluated. In combination with antiretroviral therapy, DNA vaccines based on these conserved antigens may prove to be successful in slowing or halting disease progression and improving survival rates.

## Hepatitis

Hepatitis is a viral infection of the liver cause by virus clades A-E. Hepatitis A (HAV) is contracted by the faecal-oral route and does not cause chronic liver disease. Globally, approximately 1.4 million HAV cases are reported [30]. Hepatitis B (HBV) can instigate acute and chronic liver disease and is transmitted by contact with contaminated bodily fluids. Currently over 2 billion people have been infected with the HBV virus and 350 million live with chronic infection; this is attributed to the fact that HBV virus is 50-100 times more infectious than HIV [31]. These global infection rates are extraordinarily high considering that there are safe and effective prophylactic vaccines available against HAV and HBV. Hepatitis C (HCV) is a prevalent cause of acute hepatitis and chronic liver disease, which includes cirrhosis and liver cancer. Worldwide approximately 170 million persons are chronically infected with HCV, with 4 million carriers in Europe alone and 3 -4 million persons are newly infected each year [32]. Therapeutic DNA vaccines are being developed for the two chronic infections of HBV and HCV.

To efficiently eliminate HBV infection in a chronic patient T cell-mediated immunity is crucial and a number of therapeutic DNA vaccines are in development to elicit this response (reviewed in [33]). A DNA vaccine expressing HBV envelope proteins elicited HBV-specific T cell responses in patients who were unresponsive to standard antiviral treatments; however this proliferation was only transient [34]. A later clinical study that combined a HBV DNA vaccine that consisted of most HBV genes with a lamivudine treatment regime elicited strong Type 1 T cell responses, particularly CD4+ memory T cells and had a 50% virological response rate 52 weeks after vaccination [35]. Even more promising was that of the subjects who responded to the combination therapy, two subjects continued to have undetectable viremia 3 years after the last treatment [36].

Current clinical trials involving therapeutic DNA immunization to address chronic disease state of hepatitis include two phase II trials for the combination therapy of DNA vaccination and lamivudine versus a placebo and a lamivudine treated group for chronic HBV patients. One is evaluating a double plasmid HBV DNA vaccine (ClinicalTrials.gov ID: NCT01189656) and the other is using a naked HBV DNA vaccine (ClinicalTrials.gov ID: NCT00536627). Both trials are evaluating safety and tolerance; however the first phase I trial is determining the effect of the combination on viral load and peripheral cytokines over the course of the treatment and after cessation. The second phase I/II trial is focusing on reducing viremia to undetectable levels and suppression of the clinical progression of hepatitis B.

As HCV is a highly infectious virus with an ability to rapidly mutate, clinical phase I/II trials for therapeutic DNA vaccines are rare, and of those that do get to trial stage rarely demonstrate the safety and efficacy needed to continue to phase III/IV clinical trials. For chronic HCV patients a new therapeutic vaccine known as CHRONVAC-C® is being tested for safety and tolerance in a dosage phase I/II trial (ClinicalTrials.gov ID:00563173). This is being carried out in individuals with a chronic genotype 1 infection with a viral load less than

800 IU/mL which is considered to be a low viral load. The results of this trial are due to be published in late 2011.

Hepatitis B and C are both chronic diseases that require ongoing treatment to prevent complete cirrhosis of the liver. The availability of an effective prophylaxis vaccine for HBV warrants the question of why prevention vaccination is not being implemented in areas of high prevalence. The burden this disease has on global health services is immense and successful therapeutic genetic vaccination could lessen that burden by prolonging the deterioration of the liver and reducing complications from the disease.

## Tuberculosis

Tuberculosis is a moderately infectious disease caused by the acid-fast bacterium *Mycobacterium tuberculosis* and is often associated with HIV sufferers. Over one-third of the global population is currently infected with TB and drug-resistance is an emerging problem, especially in Russia [37]. Treatment is usually by chemotherapy; however is usually applied long-term and can result in drug-resistance particularly when the chemotherapy regime is not strictly implemented. The live attenuated Bacillus Calmette-Guérin (BCG) vaccine has been administered to over 3 billion people and 100 million infants are still receiving the vaccine even though is efficacy is in question [38]. The WHO projects that if the dissemination of TB is not controlled more effectively, in twenty years more than one billion people will be newly infected and 36 million will die [37].

Protective antigens have not been elucidated for TB and the BCG vaccine is not 100% effective in adults against pulmonary TB. Various antigens delivered by genetic immunization have been tested in mice to varying levels of effectiveness with some even exacerbating the disease, reviewed in [6]. One of the earliest showed that a dramatic decrease in bacterial cell viability could be achieved using three doses of DNA vaccine after a period of chemotherapy to the point where after the third DNA vaccination no viable *M. tuberculosis* were able to be isolated from the spleens of the mice [39].

One of the most promising therapeutic vaccines to be reported is an amalgamation of DNA vaccines expressing mycobacterial heat shock protein 65 (HSP65) and human interleukin 12 (IL-12) delivered by the hemagglutinating virus of Japan (HVJ)-envelope and –liposome (HSP65 + IL-12/HVJ) [40]. When used as a prophylactic vaccine in mice and guinea pigs they showed lower levels of bacteria in the lungs, longer survival, induction of CD8+ T cells and improvement of tuberculosis lesions. When utilized as a therapeutic in murine models for both multiple drug resistant and extremely drug resistant TB, this vaccine again showed a reduction in the bacterial CFU in the lungs and prolonged survival times. This vaccine was then tested in the cynomolgus monkey model which is the best animal model for human tuberculosis. The study reported that half of monkeys in the HSP65+hIL-12/HVJ treatment group survived more than one year and two months post-infection (the termination period of the experiment) whilst all control monkeys died at 8 months. Additionally, BCG as a prime and the HSP65 + IL-12/HVJ vaccine as a booster showed a synergistic effect in the TB-infected cynomolgus monkey with 100% survival observed [40]. As this vaccine has shown efficacy across multiple animal models in both challenge trials and therapeutic trials, this indicates that it has excellent potential that should be evaluated in clinical human trials.

Another study has utilized the same strategy by creating a heat shock protein and hIL-2 fusion construct which showed excellent reduction in bacterial load in both the lungs and spleen of mice. Specifically this group observed increased CD8+ T cell levels, enhanced levels of CTL activity and elicitation of higher levels of IFN-γ, IL-2 and IgG2a compared to the traditional BCG vaccine delivered intramuscularly [41]. A difficulty with treating TB is the location of the infection deep in the lungs. One study using a DNA vaccine in a replication-incompetent recombinant adenovirus to deliver TB antigens 85A/B and mTB10.4 found that using aerosolized delivery of the vaccine to the lungs induced very strong T cell responses in macaques [42]. One study utilized the messenger RNA of *Mycobacterium leprae* HSP65 as a genetic vaccination and administered it nasally to mice to observe its efficacy. It was demonstrated that lung APCs could uptake the HSP65 mRNA and the resultant immune responses reduced the bacterial load in the lungs [43]. Both of these methods of delivery would be particularly useful for chronic TB sufferers to diminish the levels of replicating bacteria that causes TB to reactivate.

# Clinical Trials

From concept to marketing the development of a therapeutic vaccine will take many years and countless trials to demonstrate efficacy and safety. The first gene therapy experiments were conducted in the early 90s [44-46] and the first DNA vaccines to get licensure from North American regulatory bodies in July 2005 were two prophylactic DNA vaccines, West Nile-Innovator® and Apex-IHN® [47]. West Nile-Innovator® was designed by the CDC and Fort Dodge Laboratories in the USA for use in horses to protect against West Nile Virus. Apex-IHN® was designed by Novartis in Canada for farm-reared Atlantic salmon against infectious haematopoietic necrosis virus. Even though these vaccines weren't for human use, they both still underwent strict clinical trials to test for purity, potency, efficacy and safety before the United States Department of Agriculture (USDA) and the Canadian Food Inspection Agency, respectively, would grant conditional licenses [48].

Of the current clinical trials for DNA vaccines running globally 64.5% are aimed at cancer and 8% are for infectious diseases such as TB and HIV whilst the rest cover a wide variety of diseases such as cardiovascular, monogenic, neurological, ocular, etc [49]; however the majority of these trials are for prophylactic applications. Surprisingly there are only 58 therapeutic DNA vaccine clinical trials out of 608 trials listed on ClinicalTrials.gov (Figure 2). For TB there have been quite a number of prophylactic clinical trials over the last decade. However few therapeutic vaccines have reached clinical trials in patients with chronic TB. Therapeutic vaccines for other diseases have been more successful, for example for hepatitis, there have been over 10 clinical trials, for cancer there have been over 15, and for HIV over 20 clinical trials in the last decade.

# Safety Issues, Limitations and Disadvantages

The same safety issues, limitations and disadvantages that exist for traditional prophylactic DNA vaccines apply for their use as therapeutics. Numerous clinical trials have indicated that this type of subunit genetic vaccine treatment is highly unlikely to revert to a pathogenic form or cause a secondary infection in the immunocomprimised or elderly as they are inert, non-replicating and non-infectious agents which elicit potent immune responses. Ease of storage and administration, and the flexibility of this simple technology which can be monovalent or multivalent are great benefits which promote expanded use of these vaccines. However, there are some concerns with the administration of these vaccines to humans particularly to patients with chronic illnesses. There is apprehension over the risk of integration into the human genome, development of autoimmunity and the dissemination of antibiotic resistance to bacterial pathogens (reviewed in [50]).

The first prophylactic DNA vaccine to be approved for use in salmon for human consumption had to demonstrate that integration into the genome was an improbable event. Novartis demonstrated that one day post vaccination 99.5% of vaccinated salmon were negative for the plasmid [47]. In earlier work, plasmid DNA vaccine tissue dissemination and chromosomal integration was quantitatively assayed using PCR [51]. Intra-muscular vaccination of mice and guinea pigs resulted in no evidence of plasmid integration, to a sensitivity of about one copy per microgram of DNA, a value which is approximately three orders of magnitude below the spontaneous mutation frequency. This low recovery and elimination of the plasmid from the vaccinated salmon after intra-muscular injection was similar to that observed in other trials, and led the way for trials to optimize DNA vaccine administration. In the last decade, the methods of administration have changed and electroporation is a leading candidate to replace intra-muscular administration due to its higher transfection efficiency. However, while this is good news for immune responses, the more nuclei that are exposed to a therapeutic DNA plasmid the higher the probability of an integration event occurring [50].

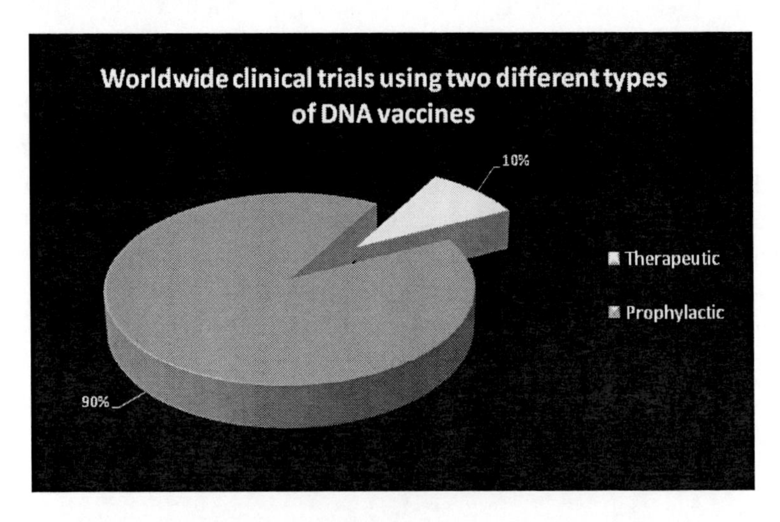


Figure 2. DNA vaccines being tested in active clinical trials as listed on ClinicalTrials.gov as of February, 2011, includes multivalent DNA vaccines for multiple conditions.

The efficacy and suitability of using electroporation as a delivery method for humans is currently being evaluated by multiple clinical phase I/II trials for both prophylactic and therapeutic purposes. The safety of electroporation administration of DNA vaccines is currently being assessed for melanoma (ClinicalTrials.gov ID: NCT00471133, NCT01138410), prostate cancer (ClinicalTrials.gov ID: NCT00859729), colorectal cancer (ClinicalTrials.gov ID: NCT01064375), hepatitis C (ClinicalTrials.gov ID: NCT00563173), HIV (ClinicalTrials.gov ID: NCT01266616) and many other diseases.

The development of autoimmunity, in primates and humans, from DNA vaccinations has not been reported from the phase I and II clinical trials and is closely monitored in participants. In fact the body of literature from these trials demonstrate that over the last fifteen years, the majority of clinical trials have shown that development of autoimmunity from prophylactic DNA vaccinations is no longer a major concern (reviewed in [52]). However as therapeutic DNA vaccines are still in their burgeoning phase and require cumulatively higher and more frequent doses, the possibility of autoimmune complications cannot be discounted. This is particularly true for the application in cancer patients where the targets of vaccination are often self-antigens that are over-expressed and the immune system is in a state of tolerance against those antigens or under immunosuppression.

Horizontal transfer of antibiotic resistance after DNA vaccination is still a major concern being investigated. In the meantime there is the added precaution in the industry to use resistance markers for antibiotics not commonly prescribed in humans. However, transfer of these plasmids to wild-type microorganisms is still such a potential issue that research is underway to remove the selective marker to improve both the manufacturing process and eliminate the risk of transference [53]. One study has recently reported that removal of the antibiotic resistance marker improves both the manufacturing process for gram-scale manufacturing of plasmids for vaccine use and the expression of the heterologous antigens being delivered [54]. If these markers can be completely eliminated from the manufacturing process then the application of these DNA vaccines can have a more widespread administration without the threat of creating more multi-drug resistant bacteria in countries such as Russia which have an existing problem.

# Conclusion

This review gives a brief snapshot of the research being conducted into the use of genetic immunization for therapy of existing diseases. This is an exciting time for DNA vaccines - there is already a therapeutic DNA vaccine available to treat cancer in dogs (ONCEPT™), and the promise is of more to come. The attractiveness of DNA as a vaccine lies in several areas; ease of manufacture (and therefore reduction in cost); ability to easily deliver multiple antigens; ability to re-program the immune system; ability to induce responses in the absence of adjuvants; and the ability to induce both the humoral and cellular arms (particularly CTLs) of the immune system. The latter point is particularly crucial, and the efficient induction of CTLs is required to provide efficacy (both prophylactic and therapeutic) against many intractable diseases.

So what of the future? It is expected that DNA vaccines will become a key strategy in the development of many new vaccines, in the same way that (for example) dendritic cell vaccines are. The challenges are to maximize the potential of the technology by building the best DNA vaccines we can, and then delivering them to the appropriate cells. This may entail strategies to improve the immunogenicity of the vaccines, particularly critical in developing vaccines against cancer, where the aim is to break tolerance against cancer antigens. We have addressed general strategies for this elsewhere (Taki et al., this volume). The delivery strategy is equally important, as while naked DNA plasmids are effective, strategies that either deliver the encoded product to dendritic cells [55-57], or utilizing particulate delivery systems to enhance uptake [58] would appear to be very productive strategies, in addition to the utilization of viral vectors.

# References

[1] Pisani, P., et al., Cancer and infection: estimates of the attributable fraction in 1990. Cancer Epidemiol. Biomarkers Prev., 1997. 6(6): p. 387-400.

[2] Speed, B., J. Kaldor, and P. Cavanagh, Guillain-Barre syndrome associated with Campylobacter jejuni enteritis. J. Infect., 1984. 8(1): p. 85-6.

[3] Bergman, P.J., et al., Development of a xenogeneic DNA vaccine program for canine malignant melanoma at the Animal Medical Center. Vaccine, 2006. 24(21): p. 4582-4585.

[4] Khan, A.S., et al., Effects of maternal plasmid GHRH treatment on offspring growth. Vaccine, 2010. 28(8): p. 1905-1910.

[5] Klebanoff, C.A., et al., Therapeutic cancer vaccines: are we there yet? Immunol. Rev., 2011. 239(1): p. 27-44.

[6] Lowrie, D.B., DNA vaccines for therapy of tuberculosis: Where are we now? Vaccine, 2006. 24(12): p. 1983-1989.

[7] Palucka, K., et al., Dendritic cells and immunity against cancer. J. Intern. Med., 2011. 269(1): p. 64-73.

[8] Buchbinder, S.P., et al., Efficacy assessment of a cell-mediated immunity HIV-1 vaccine (the Step Study): a double-blind, randomised, placebo-controlled, test-of-concept trial. Lancet, 2008. 372(9653): p. 1881-93.

[9] Liu, M.A., DNA vaccines: an historical perspective and view to the future. Immunol. Rev., 2011. 239(1): p. 62-84.

[10] Benlahrech, A., et al., Adenovirus vector vaccination induces expansion of memory CD4 T cells with a mucosal homing phenotype that are readily susceptible to HIV-1. Proc. Natl. Acad. Sci. U S A, 2009. 106(47): p. 19940-5.

[11] Wu, J., et al., Hepatitis B virus suppresses toll-like receptor-mediated innate immune responses in murine parenchymal and nonparenchymal liver cells. Hepatology, 2009. 49(4): p. 1132-40.

[12] Ulmer, J.B. and G.R. Otten, Priming of CTL responses by DNA vaccines: direct transfection of antigen presenting cells versus cross-priming. Dev. Biol. (Basel), 2000. 104: p. 9-14.

[13] Fu, T.M., et al., Priming of cytotoxic T lymphocytes by DNA vaccines: requirement for professional antigen presenting cells and evidence for antigen transfer from myocytes. Mol. Med., 1997. 3(6): p. 362-71.

[14] WHO. Cancer Fact Sheet No. 297. 2009 February 2009 [cited 2011 January 22]; Available from: http://www.who.int/mediacentre/factsheets/fs297/en/index.html.

[15] NCI. NCI Funded Research Portfolio 2009 [cited 2011 January 22]; Available from: http://www.cancer.gov/cancertopics/factsheet/NCI/research-funding.

[16] Ahmad, S., et al., Optimised electroporation mediated DNA vaccination for treatment of prostate cancer. Genet. Vaccines Ther., 2010. 8(1): p. 1.

[17] McKinney, K.A., et al., Effect of a novel DNA vaccine on angiogenesis and tumor growth in vivo. Arch. Otolaryngol. Head Neck Surg., 2010. 136(9): p. 859-64.

[18] Haller, B.K., et al., Therapeutic efficacy of a DNA vaccine targeting the endothelial tip cell antigen delta-like ligand 4 in mammary carcinoma. Oncogene, 2010. 29(30): p. 4276-86.

[19] Morse, M.A., et al., An alphavirus vector overcomes the presence of neutralizing antibodies and elevated numbers of Tregs to induce immune responses in humans with advanced cancer. J. Clin. Invest., 2010. 120(9): p. 3234-41.

[20] Collins, S.A., et al., AAV2-mediated in vivo immune gene therapy of solid tumours. Genet. Vaccines Ther., 2010. 8: p. 8.

[21] Madan, R.A., et al., Therapeutic cancer vaccines in prostate cancer: the paradox of improved survival without changes in time to progression. Oncologist, 2010. 15(9): p. 969-75.

[22] WHO. Global summary of the AIDS epidemic 2009. 2010 [cited 2011 January 22]; Available from: http://www.who.int/hiv/data/2009_global_summary.png.

[23] Montagnier, L., 25 years after HIV discovery: prospects for cure and vaccine. Virology, 2010. 397(2): p. 248-54.

[24] Sandstrom, E. and B. Wahren, Therapeutic immunisation with recombinant gp160 in HIV-1 infection: a randomised double-blind placebo-controlled trial. Nordic VAC-04 Study Group. Lancet, 1999. 353(9166): p. 1735-42.

[25] Valentin, A., et al., Repeated DNA therapeutic vaccination of chronically SIV-infected macaques provides additional virological benefit. Vaccine, 2010. 28(8): p. 1962-74.

[26] Rosenberg, E.S., et al., Safety and immunogenicity of therapeutic DNA vaccination in individuals treated with antiretroviral therapy during acute/early HIV-1 infection. PLoS One, 2010. 5(5): p. e10555.

[27] Gandhi, R.T., et al., A randomized therapeutic vaccine trial of canarypox-HIV-pulsed dendritic cells vs. canarypox-HIV alone in HIV-1-infected patients on antiretroviral therapy. Vaccine, 2009. 27(43): p. 6088-6094.

[28] Palma, P., et al., The PEDVAC trial: Preliminary data from the first therapeutic DNA vaccination in HIV-infected children. Vaccine, 2011. doi:10.1016/j. vaccine. 2010.12.058.

[29] Lapadula, G., et al., Risk of early virological failure of once-daily tenofovir-emtricitabine plus twice-daily nevirapine in antiretroviral therapy-naive HIV-infected patients. Clin. Infect. Dis., 2008. 46(7): p. 1127-9.

[30] WHO. Hepatitis A Fact sheet No. 328. 2008 May 2008 [cited 2011 January 22]; Available from: http://www.who.int/mediacentre/factsheets/fs328/en/index.html.

[31] WHO. Hepatitis B Fact Sheet No. 204. 2008 August 2008 [cited 2011 January 22]; Available from: http://www.who.int/mediacentre/factsheets/fs204/en/index.html.

[32] WHO. Global alert and Response: Hepatitis C. 2010 [cited 2011 January 22]; Available from: http://www.who.int/csr/disease/hepatitis.html.

[33] Miche, M.-L., Q. Deng, and M. Mancini-Bourgine, Therapeutic vaccines and immune-based therapies for the treatment of chronic hepatitis B: perspectives and challenges. Journal of Hepatology, 2011. In Press, Accepted Manuscript.

[34] Mancini-Bourgine, M., et al., Induction or expansion of T-cell responses by a hepatitis B DNA vaccine administered to chronic HBV carriers. Hepatology, 2004. 40(4): p. 874-82.

[35] Yang, S.H., et al., Correlation of antiviral T-cell responses with suppression of viral rebound in chronic hepatitis B carriers: a proof-of-concept study. Gene Ther., 2006. 13(14): p. 1110-7.

[36] Im, S.J., et al., Increase of Plasma IL-12/p40 Ratio Induced by the Combined Therapy of DNA Vaccine and Lamivudine Correlates with Sustained Viremia Control in CHB Carriers. Immune Netw, 2009. 9(1): p. 20-6.

[37] WHO. Tuberculosis Fact Sheet No. 104. 2010 November 2010 [cited 2011 January 22]; Available from: http://www.who.int/mediacentre/factsheets/fs104/en/.

[38] Svenson, S., et al., Towards new tuberculosis vaccines. Hum. Vaccin., 2010. 6(4): p. 309-17.

[39] Lowrie, D.B., et al., Therapy of tuberculosis in mice by DNA vaccination. Nature, 1999. 400(6741): p. 269-271.

[40] Okada, M., et al., A Novel Therapeutic and Prophylactic Vaccine (HVJ-Envelope / Hsp65 DNA + IL-12 DNA) against Tuberculosis Using the Cynomolgus Monkey Model. Procedia in Vaccinology, 2010. 2(1): p. 34-39.

[41] Changhong, S., et al., Therapeutic efficacy of a tuberculosis DNA vaccine encoding heat shock protein 65 of Mycobacterium tuberculosis and the human interleukin 2 fusion gene. Tuberculosis, 2009. 89(1): p. 54-61.

[42] Song, K., et al., Genetic immunization in the lung induces potent local and systemic immune responses. Proc. Natl. Acad. Sci. U S A, 2010. 107(51): p. 22213-8.

[43] Lorenzi, J.C., et al., Intranasal vaccination with messenger RNA as a new approach in gene therapy: use against tuberculosis. BMC Biotechnol., 2010. 10: p. 77.

[44] Wolff, J.A., et al., Direct gene transfer into mouse muscle in vivo. Science, 1990. 247(4949 Pt 1): p. 1465-8.

[45] Tang, D.C., M. DeVit, and S.A. Johnston, Genetic immunization is a simple method for eliciting an immune response. Nature, 1992. 356(6365): p. 152-4.

[46] Ulmer, J.B., et al., Heterologous protection against influenza by injection of DNA encoding a viral protein. Science, 1993. 259(5102): p. 1745-9.

[47] Salonius, K., et al., The road to licensure of a DNA vaccine. Current Opinion in Investigationaly Drugs, 2007. 8(8): p. 635-641.

[48] Tonheim, T.C., J. Bøgwald, and R.A. Dalmo, What happens to the DNA vaccine in fish? A review of current knowledge. Fish and Shellfish Immunology, 2008. 25(1-2): p. 1-18.

[49] Anon. Indications addressed by Gene Therapy Clinical Trials. 2010 June 2010 [cited 2011 January 22]; Available from: http://www.wiley.com/legacy/wileychi/ genmed/ clinical/.

[50] Faurez, F., et al., Biosafety of DNA vaccines: New generation of DNA vectors and current knowledge on the fate of plasmids after injection. Vaccine, 2010.

[51] Ledwith, B.J., et al., Plasmid DNA vaccines: assay for integration into host genomic DNA. Dev. Biol.(Basel), 2000. 104: p. 33-43.

[52] Klinman, D.M., et al., FDA guidance on prophylactic DNA vaccines: Analysis and recommendations. Vaccine, 2010. 28(16): p. 2801-2805.

[53] Mairhofer, J., et al., Marker-free plasmids for gene therapeutic applications--Lack of antibiotic resistance gene substantially improves the manufacturing process. Journal of Biotechnology, 2010. 146(3): p. 130-137.

[54] Mairhofer, J., et al., Marker-free plasmids for gene therapeutic applications--lack of antibiotic resistance gene substantially improves the manufacturing process. J. Biotechnol., 2010. 146(3): p. 130-7.

[55] Boyle, J.S., J.L. Brady, and A.M. Lew, Enhanced responses to a DNA vaccine encoding a fusion antigen that is directed to sites of immune induction. Nature, 1998. 392(6674): p. 408-11.

[56] Nchinda, G., et al., The efficacy of DNA vaccination is enhanced in mice by targeting the encoded protein to dendritic cells. J. Clin. Invest., 2008. 118(4): p. 1427-36.

[57] Rainczuk, A., et al., The protective efficacy of MSP4/5 against lethal Plasmodium chabaudi adami challenge is dependent on the type of DNA vaccine vector and vaccination protocol. Vaccine, 2003. 21(21-22): p. 3030-42.

[58] Scheerlinck, J.P. and D.L. Greenwood, Virus-sized vaccine delivery systems. Drug Discov. Today, 2008. 13(19-20): p. 882-7.

In: DNA Vaccines: Types, Advantages and Limitations
Editors: E. C. Donnelly and A. M. Dixon
ISBN 978-1-61324-444-9


*Chapter VIII*

# DNA Vaccination against Herpesvirus

***Stefano Petrini[1] and Maura Ferrari[2]***
[1]Istituto Zooprofilattico Sperimentale dell'Umbria e delle Marche, via Gaetano Salvemini, 1, 06126 Perugia, Italy
[2]Istituto Zooprofilattico Sperimentale della Lombardia e dell'Emilia-Romagna, via Antonio Bianchi 7/9, 25124 Brescia, Italy

## Abstract

Infectious diseases inflict heavy economic losses in cattle, although they are primarily preventable by vaccination. Among the different types of vaccines available, DNA immunization is one of the most recent and technologically advanced strategies to immunize against a highly prevalent infectious agent: bovine herpesvirus type 1 (BoHV-1). In the last ten years, DNA technology has offered the opportunity to retain the advantage of recombinant expression of immunodominant antigens while overcoming safety issues and logistical inconveniences inherent to live replicating vectors. Different studies on DNA vaccines against BoHV-1 have been based on expression of gD gene with a truncated trans-membrane domain. The immunizing DNA can be inoculated by conventional route (syringes) or by gene guns and with electroporation system, in order to increase the humoral and cell mediated immune responses. Different adjuvants have been inserted in DNA vaccines. However, the mechanism of action of these adjuvants is not yet completely understood and it is known that different adjuvants use different ways of estimulating the immune system.

The expressed proteins can readily go through processing and presentation via class II and class I pathways, which will result in induction of CD4+ and CD8+ T-cell responses. DNA vaccines present, howler, several inconveniences, for instance, they can stimulate immunity exclusively against proteins but not against polysaccharides; thus, at present time there is not valid practical option for polysaccharide-based-vaccines. Safety is a strong qulity of DNA vaccines: as there is no complete homology between mammalian cells and plasmid DNA occurrence of a homologous recombination is highly unlikely. However, the possibility of random integration of DNA into the host genome is a valid concern although no adverse feature of that sort has been reported so far.

# Introduction

DNA vaccines consist of a vector plasmid with a strong viral promoter, the gene of interest and a polyadenylation/transcriptional termination sequences. (Figure 1).

Vaccination is one of the most effective strategies to prevent infectious diseases in cattle [1-3]. Different types of vaccines are currently available, such us: 1) traditional vaccines (killed, live attenuated, subunit, synthetic peptides, toxoid, chimera); 2) vectored DNA; 3) bacterial and yeast carriers; 4) anti-idiotypic vaccines; 5) subunit and multi-combination vaccines; 6) adiuvants [4,5]. Although all of them can be at least partially successful, to improve both safety and efficacy of vaccination and, mainly, duration of immunity DNA immunization represents a further possible approach [6,7]. The DNA method of immunization is extremely simple as it involves the direct introduction of naked DNA into the muscle or epidermis of the recipient host by direct injection or particle bombardment. The cells of the host take up the plasmids, and under the control of some elements in the plasmid, the necessary gene is transcribed and translated to produce the immunogen that is recognized as being foreign and that can stimulate thes host cell immunity [8,9]. Thus, DNA vaccines have potentially wide-ranging applications, from vaccinations against a variety of infectious diseases to its use in cancer therapy [10-15]. DNA vaccination pioneering work was carried out by Wolff et al., (1990) who injected a plasmid encoding a reporter gene into an animal with the subsequent expression of that gene *in vivo* [16].

The plasmid is grown in bacteria (*E. coli*), purified and dissolved in a saline solution, and then simply injected into the host alone or combined with chemical adjuvants. Different studies have shown that not only was plasmid capable of expressing protein, but also that the protein produced by the plasmid DNA also induced an immune response in animals [17-21]. Furthermore, immunity stimulated by DNA vaccination results in antigen presentation through both the major histocompatibility complex class I and II pathways, and activation of humoral and cell-mediated immune responses, similar to a natural viral infection [22-25].

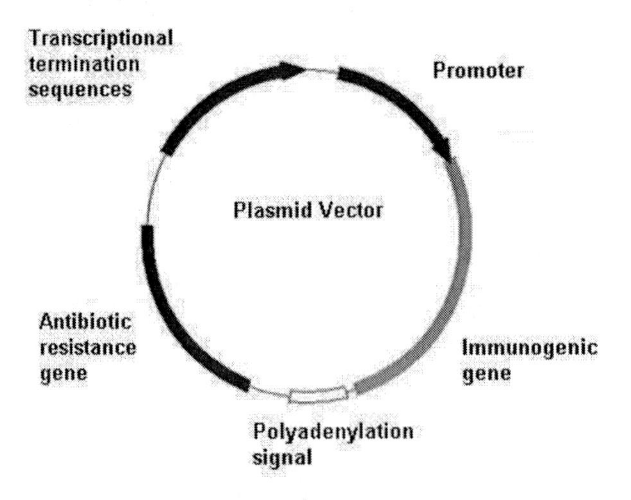


Figure 1. Representation of the contents of a Plasmid Verctor: promoter, immunogenic gene, polyadenylation signal sequence. A bacterial transcriptional termination sequences and an antibiotic resistance gene are inserted in the plasmid vector to permit growth and selection of the plasmid in bacteria.

In the past 10 years a rapid evolution of DNA immunization has taken place [26-31]. Among the different viral infectious diseases of cattle, those caused by bovine herpesvirus play a key role due to their high prevalence associated to heavy economic losses [23,32-39]. The following chapter summarizes the results obtained by bovine herpesvirus-1 DNA vaccination, the chemical adjuvants used, and the immune response obtained as a result of a DNA vaccination approach.

# Bovine Herpesvirus-1

Bovine herpesvirus-1 (BoHV-1) is a member of the subfamily, *Alphaherpesvirinae,* genus varicellovirus [40]. BoHV-1 encode for approximately 70 proteins. Most of the genes that have been so far identified are similar to those detected in other alphaherpesvirus, such as herpes simplex virus (HSV). Some proteins and glycoproteins of BoHV-1 have been shown to induce neutralizing antibody, stimulate leukocytes, and induce cytotoxic cells against viral infected cells [40]. The virus is worldwide spread and causes a variety of diseases which include: rhinotracheitis, conjunctivitis, enteritis, abortions, pustular vaginitis, fetal death, encephalitis [41,42]. Moreover, it causes reduction of milk production and infertility. BoHV-1 is one of the viruses included in the so-called "bovine respiratory diseases complex" (BRDC) [43-46] or "shipping fever" [47]. An infection sustained by this virus is frequently followed by secondary bacterial infections, in which *Mannheimia haemolytica, Haemophylus somnus,* and *Micoplasma bovis* are commonly recognized [47]. Serous to mucopurulent nasal discharge, abnormal breathing sounds, and high respiratory frequency are usually observed. Humoral and cell-mediated immunity protect against clinical signs. BoHV-1 infections can spread through a herd within a few weeks infecting all susceptible animals. The virus follows the respiratory route after contact with infected secretions or by inhalation; then it spreads into the animal by viremia and neuronal invasion. It may also be transmitted by infected frozen semen and embryos. Conjunctivitis with an ocular discharge is frequently observed in the first stage of the disease. The infection induces a necrosis of the epithelium of the muzzle, nose, nasopharynx, trachea, and the bronchi. After the infection, the virus migrates to trigeminal or sacral ganglia where it remains in latent phase of infection. Reactivation from latency and virus shedding may then occur at any time after acute pahse of infection. Factors responsible for reactivation are diverse; including among them, stress, transport distress, weather changing, viral co-infection, and overcrowding, amongst the most conspicuous that can be recognized [48].

# DNA Vaccine against Bovine Herpesvirus-1

Through the DNA immunization approach it can be demonstrated that inoculation of the plasmid encoding for a viral gene into a mouse resulted in gene expression *in vivo.* The use of DNA vaccination for BoHV-1 has shown promising results in mice [24,33,35]. Actually, it is very easy to induce immune responses in mice, but it is difficult to repeat the same experiments and to obtained identical success with these gene constructions in large animals [8,35,48].

Several BoHV-1 glycoproteins, such as gB, gC, gD and gD with a truncated trans-membrane domain (tgD) are effective immunogens and partially protect the calves from infection upon challenge with virulent virus [23,24,36,49-52].

BoHV-1 gB is involved in attachment and penetration of BoHV-1virus into cells [53-57], involving fusion between the virion envelope and the cellular plasma membrane, as well as direct transmission of infection from primary infected to non-infected cells [57-59]. While BoHV-1 gC is a major viral attachment protein [50,60] it interacts with a cellular heparin–like moiety. It induces both neutralizing antibodies and cytotoxic T-lymphocyte responses in animals immunized or infected with BoHV-1 [60,61]. The gD has been shown to be a major target for CD8+ cytotoxic lymphocytes [61] and for neutralizing antibodies [32,49,62,63].

Different plasmids were constructed by cloning the BoHV-1 gene encoding gB, gC, gD or tgD, under the control of enhancer/early different promoters and with polyadenylation signal.

## Route of Immunization

Although DNA vaccinations are mainly delivered intramuscularly, several other inoculation routes have been examined (Table 1). Complete protection against influenza virus has been achieved by either intradermal, intranasal, intraperitoneal, or intravenous routes carried out in laboratory animals [64]. Further studies showed an immune response in animal inoculated by intravenous, sublingual, intradermal, intraperitoneal and vaginal routes [17,22,23,29,36,65]. When orally injected, a DNA vaccine can also elicit immune response if DNA is protected from degradation [66,67]. Finally, DNA vaccine has also been injected in calves by intramuscular route in the retroauricolar region [23].

## Chemical Adjuvants

Adjuvants in DNA vaccines have been used with the purpose of increasing immune responses. These adjuvants can improve extent, localization or duration of plasmid DNA expression, or target its delivery into specific cells. The mechanism of action of most adjuvants is poorly understood and can occur in different ways, such us increasing the duration of immune response, activating CTLs, enhancing avidity and specificity of antibody responses, preventing or reducing antigen competition in combined vaccines, and increasing the potency of antigens.

Adjuvants that enhance the immune response may be represented by vaccine delivery vehicles such as lipid particles and particulate adjuvants, or may consist of immunostimulatory molecules such as, CpG (Figure 2) cytokines, LPS, UbilacI chimeric proteins (Figure 3) and TLR agonists [68-73].

Alum salts are the most used adjuvants for traditional vaccines [74]. Formulation of DNA vaccine in alum salts enhanced humoral immune responses in rodents and primates and did not alter the expression of immunogen [74].

**Table 1. Summary of DNA vaccines used in different studies conducted in Italy**

| Agent | Antigen | Route | Dose | Adjuvant | Immunological Response | Protection | Reference |
|---|---|---|---|---|---|---|---|
| BoHV-1 | tgD | i.m. | 500 μg | None | Partial | Partial | Castrucci *et al.*, 2004 |
| BoHV-1 | tgD | i.n. | 500 μg | None | N.D. | N.D. | Castrucci *et al.*, 2004 |
| BoHV-1 | tgD | i.d. | 500 μg | None | N.D. | N.D. | Castrucci *et al.*, 2004 |
| BoHV-1 | tgD | i.m. | 1000 μg | None | N.D. | N.D. | Petrini *et al.*, 2011 |
| BoHV-1 | tgD | i.m. | 1000 μg | 48 motif of CpG (500 μg) | Partial | Partial | Petrini *et al.*, 2011 |
| BoHV-1 | tgD | i.m. | 1000 μg | Ubilac Protein | N.D. | N.D. | Petrini *et al.*, 2011 |
| BoHV-1 | tgD | i.m. | 1000 μg | 48 motif of CpG (500 μg)+ Ubilac protein | N.D. | N.D. | Petrini *et al.*, 2011 |
| BoHV-1 | gB | i.m.,s.c., i.n. | 100 μg | 20 μg Immuneasy adjuvant | N.D. | N.D. | Caselli *et al.*, 2005 |
| BoHV-1 | tgB | i.m.,s.c., i.n. | 100 μg | 20 μg Immuneasy adjuvant | N.D. | N.D. | Caselli *et al.*, 2005 |
| BoHV-1 | gD | i.m.,s.c., i.n. | 100 μg | 20 μg Immuneasy adjuvant | N.D. | N.D. | Caselli *et al.*, 2005 |
| BoHV-1 | tgD | i.m.,s.c., i.n. | 100 μg | 20 μg Immuneasy adjuvant | N.D. | N.D. | Caselli *et al.*, 2005 |
| BoHV-1 | gB+gD | i.m.,s.c., i.n. | 100 μg | 20 μg Immuneasy adjuvant | Partial | Partial | Caselli *et al.*, 2005 |
| BoHV-1 | tgB+tgD | i.m.,s.c., i.n. | 100 μg | 20 μg Immuneasy adjuvant | Partial | Partial | Caselli *et al.*, 2005 |

BoHV-1, Bovine herpesvirus-1; tgD, truncated glycoprotein D of BoHV-1; gB, glycoprotein B of BoHV-1; tgB; truncated glycoprotein B of BoHV-1; gD, glycoprotein D of BoHV-1; i.m., intramuscular route; i.n., intranasal route; i.d., intradermal route; s.c., subcutaneous route; CpG, cytidine-phosphate-guanosine; Immuneasy adjuvant, DNA adjuvant; N.D., not determined.

Different cytokines have been evaluated as possible adjuvants for DNA vaccines and regulate the stimulating T cells or direct the infected cells. Investigate cytokines include: IL-2; IL-7; IL-12 and IL-15 [75-77].

Cytokine MCP-1; MIP-1α, Flt-3L, and IL-8 [75,78-79] can also activate antigen presenting cells (APCs). Moreover, GM-CSF can recruit APCs and macrophages [80-82], but GM-CSF has not been shown to enhance MHC class I cellular immune response after a vaccine injected by intramuscular route [83].

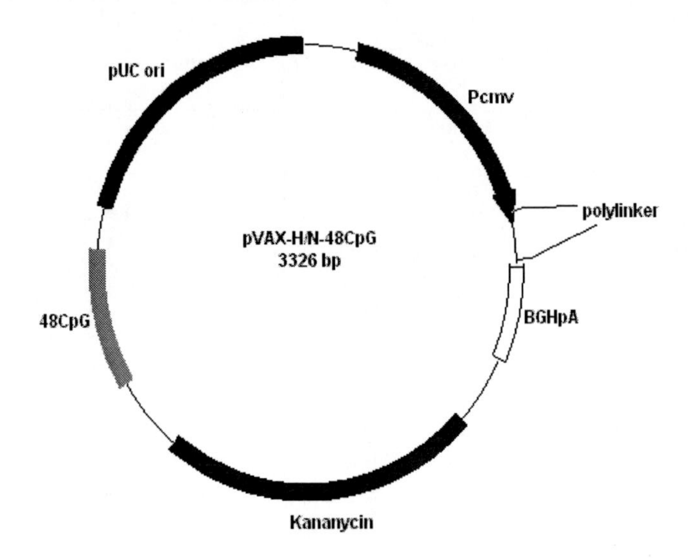


Figure 2. Representation of the contents of a Plasmid Vector: promoter (Pcmv), immunogenic gene (48 motif of CpG), polyadenylation signal sequence (BGHpA). A bacterial transcriptional termination sequence (pUO ori) and an antibiotic resistance gene (Kanamycin) are inserted in the plasmid vector to permit growth and selection of the plasmid in bacteria.

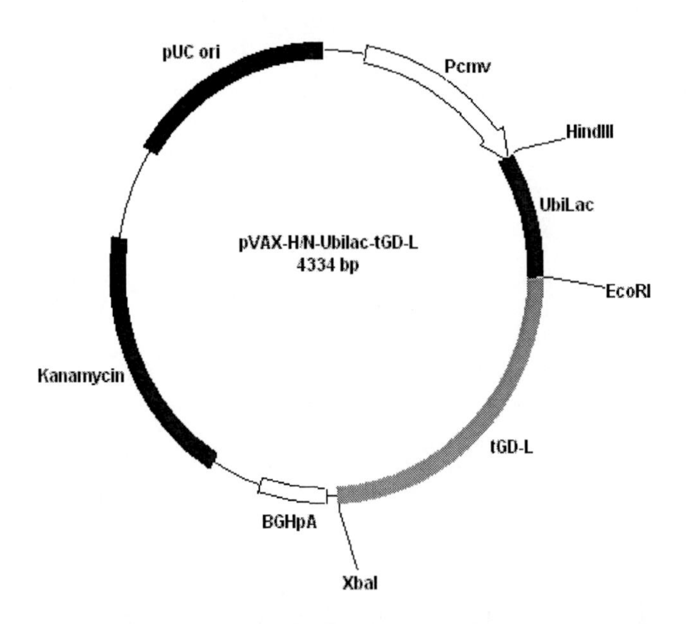


Figure 3. Representation of the contents of a Plasmid Vector: promoter (Pcmv), immunogenic genes (UbiLac and tgD), polyadenylation signal sequence (BGHpA). A bacterial transcriptional termination sequences (pUC ori) and an antibiotic resistance gene (Kanamycin) are inserted in the plasmid vector to permit growth and selection of the plasmid in bacteria.

Recently [23,24,84] the incorporation of CpG oligodeoxynucleotides (ODN) has been tested in order to evaluate its ability to enhance immune responses against viral antigens.

*In vitro* studies have shown that ODN containing CpG motifs can active macrophages as demonstrated by secretion of IL-12, TNF-α, INF-α and IFN-β [85,86]. CpG are also mitogenic for B cells and can stimulate dendritic cells and also indirectly induce NK cells to secrete IFN-γ and enhance their lytic activity [87-89].

When co-administered with other adjuvants (i.e. Freund's incomplete), CpG can increase the immunogenic potential to a level similar to the immune response induced by Freund's complete adjuvant [90]. Different studies have shown that CpG injected with oil-based adjuvants can stimulate immune responses higher than those induced by CpG in alum salts [91].

CpG motifs have been included in a commercial Hepatitis B Virus (HBV) vaccine, due to their ability to increase its immunogenicity [92].

# Animal Response Following DNA Vaccination

Different factors are involved in a response to DNA vaccination. Among these factors, the route of injection plays a key role. The antibody response induced by DNA vaccination was first demonstrated in mice after particle bombardment of gold beads coated with DNA encoding for human growth hormone and human α-1 antitrypsin [93].

Antibodies against viral proteins were demonstrated after inoculation of BoHV-1 virus glycoproteins [1,2,28] as well as against other viral infections [20,94,95]. In some cases, these antibodies were capable to protect against challenge with the virulent virus, indicating that the antigens expressed *in vivo,* after DNA vaccination can assume a natural structure with the conformation of the epitopes. In other studies, the injected DNA vaccines [23,36], were not able to stimulate a measurable immune response. Generally, the antibody response induced by DNA vaccines are mainly of the IgG isotype, but IgM and IgA have also been detected together with low levels of neutralizing antibodies [23,36]. Several studies have focused on the type of immune response induced upon DNA immunization, and the results have shown that the method of delivery plays a relevant role in eliciting the immune response toward Th1 or Th2 type response [96,97]. Generally, a DNA vaccine injected by intramuscular or intradermal routes favors Th1 type response whereas inoculation by electroporation may tend to favor Th2 response [98]. Moreover, several studies have been conducted with the aim to detect a relationship between antigen expression and immune response. Babiuk et al., (1999) have shown that different antigenic forms of BoHV-1 gD had induced both humoral and cell-mediated immune response [8]. The cytosolic and membrane-anchored forms favored antibodies of the $IgG_{2a}$ isotype, while the secreted form favored $IgG_1$ synthesis. The same authors demonstrated that the cell compartment of origin of the antigen can be responsible for the stimulation of a different immune response type. Mice injected with DNA vaccine encoding *β*-galactosidase showed $IgG_{2a}$ antibody isotype secretion and $CD4^+$, T cell secretion of IFN-γ, but not IL-4 and IL-5. In contrast, mice given protein in saline solution, produced $IgG_1$ and IgE, CD4+ T cell secretion of IL-4 and Il-5, but not IFN-γ.

These findings seem to indicate that DNA vaccination given by intramuscular route stimulates a predominant Th1 response, whereas protein immunization induced Th2 response to same antigen [99].

## Acknowledgments

The authors are grateful to Prof. Fernando A. Osorio, School of of Veterinary Medicine & Biomedical Sciences, University of Nebraska-Lincoln (USA) for providing a critical review of this chapter.

## References

[1] Babiuk L.A., Rouse B.T. (1996). Herpesvirus vaccines. Advance Drug Delivery Reviews, 21, 63-67.

[2] Van Drunen Little-van den Hurk S. (2006). Rational and perspectives on the success of vaccination against bovine herpesvirus-1. Veterinary Microbiology, 113, 275-282.

[3] Castrucci G., Frigeri F., Salvatori D., Ferrari M., Lo Dico M., Rotola A., Sardonici Q., Petrini S., Cassai E. (2002). A study on latency in calves by five vaccines against bovine herpesvirus-1 infection. Comparative Immunology Microbiology and Infectious Diseases, 25, 205-215.

[4] Carter P.B., Carmichael L.E. (2003). Modern veterinary vaccines and the Shaman's apprentice. Comparative Immunology Microbiology and Infectious Diseases, 26, 389-400.

[5] Castrucci G., Ferrari M., Salvatori D., Sardonini Q., Frigeri F., Petrini S., Lo Dico M., Marchini C., Rotola A., Amici A., Provinciali M., Tosini A., Angelini R., Cassai E. (2005). Vaccination trias against Bovine herpesvirus-1. Veterinary Research Communications, 29(2), 229-231.

[6] Donelly J.J., Ulmer J.B., Liu M.A. (1994). Immunization with DNA. Journal of Immunological Methods, 176, 145-152.

[7] Donelly J.J., Ulmer J.B., Shiver J.W., Liu M.A. (1997). DNA vaccines. Annual Review Immunology, 15, 617-648.

[8] Babiuk L.A., Lewis J., Van den Hurk S., Braun R. (1999). DNA immunization: Present and future. Advances in Veterinary Medicine, 41, 163-179.

[9] Petrini S., Ferrari M., Vincenzetti S., Vita A., Amici A., Ramadori G. An Immunoenzyme Linked Assay (ELISA) for the detection of Antibodies to Truncated Glycoprotein D (tgD) of Bovine Herpesvirus-1. Veterinary Research Communications 2006;30:257-259.

[10] Makino M., Uemura N., Moroda M., Kikumura A., Piao L.X., Mohamed R.M., Aosai F. (2010). Innate Immunity in DNA vaccine with Toxoplasma gondii heat shock protein 70 gene that induces DC activation and Th1 polarization. Vaccine, In press, DOI

[11] Adoga M.P., Pennap G., Okande B.O., Mairiga J.P., Pechulano S., Agwale S.M. (2010). Evaluation of a recombinant DNA hepatitis B vaccine in a vaccinated Nigerian population. The Journal of infection in Developing Countries, 4(11), 740-744.

[12] Eschenburg G., Stermann A., Preissner R., Meyer H.A., Lode H.N. (2010). DNA vaccination: Using the patient's Immune System to Overcome Cancer. Clinical and Developmental Immunology, 16, 1-14.

[13] Keita U., Kawakami S., Suzuky R., Maruyama K., Yamashita F., Haschida M. (2011). Suppression of melanoma growth and metastasis by DNA vaccination using a ultrasound-responsive and mannose-modified gene carrier. Molecular Pharmaceutics, In Press, DOI: 10.1021/mp100369n.

[14] Ferrari M., Petrini S., Ramadori G., Gregori A., Corradi A., Villa R., Lombardi G., Borghetti P., Bottarelli E., Bonati L., Amici A. (2006). Inoculation of two plasmid types encoding for PRRS virus ORF 4 and ORF 5 in pigs. International Pig Veterinary Society, (IPVS) 2006:26.

[15] Ferrari M., Petrini S., Ramadori G., Corradi A., Lombardi G., Borghetti P., Villa R., Bottarelli E., Amici A. (2006). Evalutation of efficacy of DNA vaccination against bovine herpesvirus-1 in calves. World Buiatrics Congress, 1,96.

[16] Wolff J.A., Malone R.W., Williams P., Acscadi G., Jani A., Felgner P.L. (1990). Direct gene transfer into mouse muscle in vivo. Scince, 242, 1465-1468.

[17] De Rose R., Tennet J., McWaters P., Chaplin P.J., Wood P.R., Kimpton W., Cahill R., Scheerlinec J-P. Y. (2002). Efficacy of DNA vaccination by different routes of immunization in sheep. Veterinary Immunology and Immunopathology, 90, 55-63.

[18] Ruitenberg K.M., Walker C., Wellington J.E., Love D.N., Whalley J.M. (1999). Potential of DNA vaccination for equine herpesvirus 1. Veterinary Microbiology, 68, 35-48.

[19] Schop G.L., Elankumaran S., Hecker R.A. (2002). DNA vaccination in the avian. Veterinary Immunology and Immunopathology, 89, 1-12.

[20] Spatz S., Maes R.K. (1993). Immunological characterization of the feline herpesvirus-1 glycoprotein B and analysis of its deduced amino acid sequences. Virology, 197, 125-136.

[21] Le Potier M.F., Monteil M., Houdayer C., Eliot M. (1997). Study of the delivery of the gD gene of pseudorabies virus to one-day-old piglets by adenovirus or plasmid DNA as ways to by-pass the inhibition of immune response by colostral antibodies. Veterinary Microbiology, 55, 75-80.

[22] Soboll G., Hussey S.B., Whalley J.M., Allen G.P., Koen M.T., Santucci N., Fraser D.G., Macklin M.D., Swain W.F., Lunn D.P. (2006). Antibody and cellular immune responses following DNA vaccination and EHV-1 infection of ponies. Veterinary Immunology and Immunopathology, 111, 81-95.

[23] Petrini S., Ramadori G., Corradi A., Borghetti P., Lombardi G., Villa R., Bottarelli E., Guercio A., Amici A., Ferrari M. (2011). Evaluation of safety and efficacy of DNA vaccines against bovine herpesvirus-1 (BoHV-1) in calves. Comparative Immunology, Microbiology and Infectious Diseases, 34, 3-10.

[24] Caselli E., Boni M., Di Luca D., Salvatori D., Vita A., Cassai E. (2005). A combined bovine herpesvirus 1 gB-gD DNA vaccines induces immune response

in mice. Comparative Immunology, Microbiology and Infectious Diseases, 28, 155-166.
[25] Huang Y., Babiuk L.A., Van Drunen Little-van den Hurk S. (2005). Immunization with a bovine herpesvirus 1 glycoprotein B DNA vaccine induces cytotoxic T-lymphocyte responses in mice and cattle. Journal of General Virology, 86, 887-898.
[26] Watts A.M., Kennedy R.C. (1999). DNA vaccination strategies against infectious diseases. International Journal for Parasitology, 29, 1149-1163.
[27] Krishnan B.R. (2000). Current status of DNA vaccines in veterinary medicine. Advance Drug Delivery Reviews, 43, 3-11.
[28] Kraehenbuhl J.P. (2001). Mucosa target DNA vaccination. Trends in Immunology, 22 (12), 646-648.
[29] Gerdts V., Tsang C., Griebel P.J., Babiuk L.A. (2004). DNA vaccination in utero: a new approach to induce protective immunity in the newborn. Vaccine, 22, 1717-1727.
[30] Huygen K. (2005). Plasmid DNA vaccination. Microbes and Infection, 7, 932-938.
[31] Faurez F., Dory D., Le Moigne V., Gravier R., Jestin A. (2010). Biosafety of DNA vaccines: New generation of DNA vectors and current knowledge on the fate of plasmid after injected. Vaccine, 28, 3888-3895.
[32] Van Drunen Little-van den Hurk S., Parker M.D., Massie B., Van den Hurk B., Harland J.V., Babiuk L.A., Zamb T.J. (1993). Protection of cattle from BHV-1 infection by immunization with recombinant glycoprotein gIV. Vaccine, 11, 25-35.
[33] Cox G.J.M., Zamb T.J., Babiuk L.A. (1993). Bovine herpesvirus 1: Immune responses in mice and cattle injected with plasmid DNA. Journal of Virology, 67, 5664-5667.
[34] Tikooo S.K., Fitzpatrick D.R., Babiuk L.A., Zamb T.J. (1990). Molecular coloning, sequencing and expression of functional bovine herpesvirus 1 glycoprotein gIV in transfected bovine cells. Jouranl of Virology, 64, 5132-5142.
[35] Babiuk L.A., Lewis P.J., Cox G., Van Drunen Little-van den Hurk S., Baca-Estrada M., Tikoo S.K. (1995). DNA immunization with bovine herpesvirus-1 genes. Series. New York Academy of Sciences. New York, pp. 50-60.
[36] Castrucci G., Ferrari M., Marchini C., Salvatori D., Provinciali M., Tosini A., Petrini S., Sardonini Q., Lo Dico M., Frigeri F., Amici A. (2004). Immunization against bovine herpesvirus-1 infection. Preliminary tests in calves with a DNA vaccine. Comparative Immunology, Microbiology and Infectious Diseases, 27, 171-179.
[37] Toussaint J.F., Coen L., Letellier C., Dispas M., Gillet L., Vanderplasschen A., Kerkhofs P. (2005). Veterinary Research, 36, 529-544.
[38] Gupta P.K., Saini M., Gupta L.K., Rao V.D.P., Bandyopadhyay S.K., Butchaiah G., Garg G.K., Garg S.K. (2001). Induction of immune responses in cattle with a DNA vaccine encoding glycoprotein C of bovine herpesvirus-1. Veterinary Microbiology, 78, 293-305.

[39] Zheng C., Babiuk L.A., Van drunen Little-van den Hurk S. (2005). Bovine herpesvirus 1 VP 22 enhances the efficacy of DNA vaccine in cattle. Journal of Virology, 79 (3), 1948-1953.

[40] Roizman B., Pellet P.E. The family of herpesviridae: a brief introduction, vol. 2. Philadelphia, PA: Lippincott Williams and Wilkins; 2001. p. 2381-2397.

[41] Nandi S., Kumar M., Manohar M., Chauhan R.S. (2009). Bovine herpesvirus infections in cattle. Animal Health Research Reviews, 10, 85-98.

[42] Muylkens B., Thiry J., Kirten P., Schynts F., Thiry E. (2007). Bovine herpesvirus 1 infection and infectious bovine rhinotracheitis. Veterinary Research, 38, 181-209.

[43] Johnes C., Chowdhury S. (2010). Bovine Herpesvirus Type 1 (BHV-1) is an important cofactor in the bovine respiratory disease complex. Veterinary Clinic Food Animal, 26, 303-321.

[44] Petrini S., Paniccià M., Briscolini S., Cucco L., Ferrari M., Filippini G., Pezzetti G. (2008). Monitor Project : Evidence of some respiratory virus isolation from Marchigiana breeding farms with respiratory disorders, note 2. XXV World Buiatrics Congress, 2,102.

[45] Paniccià M., Filippini G., Duranti A., Ciuti F., Dorinzi A., Papa P., Mangili P., Petrini S.(2008) Monitor Project 2: Serological survey in Marchigiana breeding farms, note 1. XXV World Buiatrics Congrss, 2,103.

[46] Clinton J., Chouwdhury S. (2010). Bovine herpesvirus type 1 (BHV-1) is an important cofactor in the Bovine Respiratory Disease Complex. Veterinary Clinics of North America: Food Animal Practice, 26(2), 303-321.

[47] Valarcherch J.F., Hagglund S. (2006). Viral respiratory infections in cattle. 24th World Buiatrics Congress, In Proceedings, pp. 384-397.

[48] Tikoo S.K., Campos M., Babiuk L.A. (1995). Bovine herpesvirus 1 (BHV-1): Biology, Pathogenesis and Control. Advances Virus Research, 45, 191-223.

[49] Babiuk L.A., L'Italien J., Van Drunen Littel-van den Hurk S., Zamb T., Lawman J.P., Hughes G., Gifford G.A. (1987). Protection of cattle from bovine herpesvirus type 1 (BHV-1) infection by immunization with individual viral glycoproteins. Virology, 159(1), 57-66.

[50] Gupta P.K., Saini M., Bandyopadhayay S.K., Garg S.K. (1998). Bovine herpesvirus type 1 glycoprotein C expression in MDBK cell and its reactivity in enzyme-linked immunosorbent assay. Acta Virology, 42, 397-400.

[51] Israel B.A., Herber R., Gao Y., Letchworth G.J. (1992). Induction of mucosal barrier to bovine herpesvirus 1 replication in cattle. Virology, 88(1), 256-264.

[52] Van Drunen Little van den Hurk S., Braun R.P., Lewis P.J., Karvonen B.C., Baca-Estrada M.E., Snider M., McCartney D., Watts T., Babiuk L.A. (1998). Intradermal immunization with a bovine herpesvirus-1 DNA vaccine induces protective immunity in cattle. Journal of General Virology, 79, 831-839.

[53] Li Y., Van Drunen Little-van den Hurk S., Liang X., Babiuk L.A. (1996). Production and characterization of bovine herpesvirus 1 glycoprotein B ectodomain derivates in a hsp70A gene promoter-based expression system. Archives Virology, 141, 2019-2029.

[54] Li Y., Van Drunen Little-van den Hurk S., Liang X., Babiuk L.A. (1997). Functional analysis of the transmembrane anchor region of bovine herpesvirus 1 glycoprotein gB. Virology, 228, 39-54.

[55] Liang X.P., Babiuk L.A., Van Drunen Little-van den Hurk S., Fitzpatrick D.R., Zamb T.J.(1991). Bovine herpesvirus 1 attachment to permissive cells is mediated by its major glycoproteins gI, gIII and gIV. Journal of Virology, 65, 1124-1132.

[56] Baranowsky E., Duboisson J., Van Drunen Little-van den Hurk S., Babiuk L.A., Micheal A., Pastoret P.P., (1995). Synthesis and processing of bovine herpesvirus-1 glycoprotein H. Virology, 206(1), 651-614.

[57] Fuller A.O., Lee W.C. (1992). Herpes simplex virus type 1 entry through a cascade of virus-cell interactions requires different roles of gD and gH in penetration. Journal of Virology, 66, 5002-5012.

[58] Kuhn J.E., Kramer M.D., Willembacher W., Wieland U., Lorentzen E.U., Braun R.W. (1990). Identification of herpes simplex virus type 1 glycoproteins interacting with the cell surface. Journal of Virology, 64, 2491-2497.

[59] Meithke A., Keil G.M., Weilland F., Mettenleiter T.C. (1995). Undirectional complementation between glycoprotein B homologues of pseudorabies virus and bovine herpesvirus 1 is determined by the carboxy-terminal part of the molecule. Journal of general Virology, 76, 1623-1635.

[60] Okazaki K., Honda E., Kono Y. (1994). Expression of bovine herpesvirus 1 glycoprotein gIII by a recombinant baculovirus in insect cells. Journal of General Virology, 75, 901-904.

[61] Denis M., Slaoui M., Keil G., Babiuk L.A., Ernst E., Pastoret P.P., Thiry E. (1993). Identification of different target glycoproteins for bovine herpes virus type 1-specific cytotoxic T lymphocytes depending on the method of in vitro stimulation. Immunology, 78, 7-13.

[62] Oliveira S.C., Harms J.S., Rosinha G.M., Rodarte R.S., Rech E.L., Splitter G.A. (2000). Biolistic-mediated gene transfer using the bovine herpesvirus-1 glycoprotein D is an effective delivery system to induce neutralizing antibodies in its natural host. Journal of Immunological Methods, 245, 109-118.

[63] Van Drunen Little-vanden Hurk S., Gifford G.A., Babiuk L.A. (1990). Epitope specificity of the protective immune response induced by individual bovine herpesvirus-1 glycoproteins. Vaccine, 8, 358-368.

[64] Fynan E.F., Webster R.G., Fuller D.H., Haynes J.R., Santoro J.S., Robinson H.L. (1993). DNA vaccines: Protective immunizations by parenteral, mucosal, and gene gun inoculations, The Proceedings of the National Academy of Sciences, USA, 90, 11478-11482.

[65] McLuskie M.J., Brazolot Milliam C.L., Gramzinski R.A. (1999). Route and method of delivery of DNA vaccine influence immune responses in mice and non-human primates. Molecular Medicine, 5(5),287-300.

[66] Chen S.C., Jones D.H., Fynan E.F., Farrar G.H., Clegg J.C., Greenberg H.B.,Hermann J.E. (1998). Protective immunity induced by oral immunization with a rotavirus DNA vaccine encapsulated in microparticles. Jouranl of Virology, 72(7), 5757-5761

[67] Jones D.H., Partidos C.D., Steward M.W., Farrar G.H. (1997). Oral delivery of poly(lactide-co-glycolide) encapsulated vaccines. Behring Institute Mitteilungen, 220-228.

[68] Pelegrì C., Rodriguez M., Serra M., Castellote C., Catsell M., French A. (1997). Decrease on CD4+ CD45RC+ T cell after an effective treatment of adjuvant arthrtritis with anti-CD4 monoclonal antibody. Immunology Letters, 56, 486.

[69] Ju Y.B., Wang B., Zhang C., Wang M. (2006). Expression of bovine IFN-γ and Foot and Mouth disease VP1 antigen in P. pastoris and their effects on mouse immune response to FMD antigens. Vaccine, 24(1), 82-89.

[70] Nichani A.K., Mena A., Popowych Y., Dent D., Townsend H.G.G., Mutwiri G.K., Hecker R., Babiuk L.A., Griebel P.J. (2004). In vivo immunostimulatory effects of CpG oligodeoxynucleotide in cattle and sheep. Veterinary Immunology and Immunopathology, 98(1-2), 17-29.

[71] Nichani A.K., Kaushik R.S., Mena A., Popowych Y., Dent D., Townsend H.G.G., Mutwiri G., Hecker R., Babiuk L.A., Griebel P.J. (2004). CpG oligodeoxynucleotides induction of antiviral effector molecules in sheep. Cellular Immunology, 227(1), 24-37.

[72] Mapletoft J.W., Oumouna M., Townsend H.G., Gomis S., Babiuk L.A., Van Drunen Littel-van den Hurk S. (2006). Formulation with CpG oligodeoxynucleotides increases cellular immunity. Virology, 353(2), 316-323.

[73] Shums H. (2005). Recent developments in veterinary vaccinology. The veterinary Journal, 170, 289-299.

[74] Ulmer J.B., DeWitt C.M., Chastain M., Friedman A., Donnelly J.J., McClements W.L., Caulfield M.J., Bohannon K.E., Volkin D.B., Evans R.K. (1999). Enhancement of DNA. Vaccine, 18(1-2), 18-28.

[75] Kim J.J., Trivedi N.N., Nottingham L.K., Morrison L., Tsai A., Hu Y., Manhlingman S., Dang K., Ahn L., Doyle N.K., Wilson D.M., Chattergoon M.A., Chaalian A., Bojer J.D., Agadjanin M.G. Weiner D.B. (1998). Modulation of amplitude and direction of in vivo immune responses by co-administration of cytokine gene expression cassettes with DNA immunogens. European Journal of Immunology, 28,1089-1103.

[76] Xin K.Q., Hamajima K., Sasaki S., Tsuji T., Watabe S., Okada E., Okada K. (1999). IL-15 expression plasmide enhances cell-mediated immunity induced by an HIV-1 DNA vaccine. Vaccine, 17, 858-866.

[77] Seaman M.S., Peyerl F.W., Jackson S.S., Lifton M.A., Gorgone D.A., Schmitz J.E., Letvin N.L. (2004). Subset of memory cytotoxic T lymphocytes elicited by vaccination influence the efficiency of secondary expansion in vivo. Journal of Virology, 78, 206-215.

[78] Sin J., Kim J.J., Pachuk C., Satishchandran C, Weiner D.B. (2000). DNA vaccines encoding interleukin-8 and RATES enhance antigen-specific Th1-type CD4(+) T-cell-mediated protective immunity against herpesvirus simplex virus type 2 in vivo. Journal of Virology, 74, 11173-11180.

[79] Barouch D.H., McKay P.F., Sumida S.M., Santra S., Jackson S.S., Gorgone D.A., Lifton M.A., Chakrabarti B.K, Xu L., Nabel G.J., Letvin N.L. (2003). Plasmid chemokines and colony-stimulating factors enhance the immunogenicity of DNA

priming-viral vector boosting human immunodeficiency virus type 1 vaccines. Journal of Virology, 77, 8729-8735.
[80] McKay P.F., Barouch D.H., Santra S., Sumida S.M., Jackson S.S., Gorgone D.A., Lifton M.A., Letvin N.L. (2004). Recruiment of different subset of antigen-presenting cells selectively modulates DNA vaccine-elicited CD4+ and CD8+ T lymphocyte responses. European Journal of Immunology, 34, 1011-1020.
[81] Chang D.Z., Lomazow W., Joy Somberg C., Stan R., Perales M.A. (2004). Granulocyte-macrophage colony stimulating factor: an adjuvant for cancer vaccines. Hematology, 9, 207-215.
[82] Brave A., Ljungberg K., Boberg A., Rollman E., Isaguliants M., Lund-gren B., Blomberg B., Hinkula J., Wharen B. (2005). Multigene/multisubtype HIV-1 vaccine induces potent cellular and humoral immune responses by needle-free intradermal delivery. Molecular Therapy, 12, 1197-1205.
[83] Barouch D.H., Santra S., Tenner-Racz K., Racz P., Kuroda M.J., Schmitz J.E., Jackson S.S., Lifton M.A., Freed D.C., Perry H.C., Davies M.E., Shiver J.W., Letvin N.L. (2002). Potent CD4+ T cell responses elicited by a bicistronic HIV-1 DNA vaccine expressing gp120 and GM-CSF. Journal of Immunology, 168, 562-568.
[84] Rankin R., Pontarollo R., Ioannou X., Krieg A.M., Hecker R., Babiuk L.A., Van Drunen-Little van den Hurk S. (2001). Identification of a stimulatory CpG motif for veterinary and laboratory species demonstrated that sequence recognition is highly conserved. Antisense and Nucleic Acid Drug Development, 11, 333-340.
[85] Kliman D.M., Ishii K.J., Verthelyi D. (2000). CpG DNA augments the immunogenicity of plasmid DNA vaccines. Current Topics in Microbiology and Immunology, 247, 131-142.
[86] Suen Y., Lee S.M., Qian J., Van de Ven C., Cairo M.S. (1998). Dysregulation of lymphokine production in the neonate and its impact on neonatal cell mediated immunity. Vaccine, 16(14-15), 1369-1377.
[87] Krieg A.M., Yi A.K., Hartman G. (1999). Mechanisms and therapeutic applications of immune stimulatory CpG DNA. Pharmacology Therapy, 84, 113-120.
[88] Rakin R., Pontarollo R., Ioannoux X., Krieg A.M., Hecker R., Babiuk L.A., Van Drunen Little-van den Hurk S. (2001). CpG motif identification for veterinary and laboratory species demonstrated that sequences recognition is highly conserved. Antisense and Nucleic Acid Drugs Development, 11, 333-340.
[89] Pontarollo R.A., Babiuk L.A., Hecker R., Van Drunen Little-van den Hurk S. (2002). Augmentation of cellular immune responses to bovine herpesvirus-1 glycoprotein D by vaccination with CpG-enhanced plasmid vectors. Journal of General Virology, 83, 2973-2981.
[90] Weeranta R.D., McKluskie M.J., Xu Y., Davis H.L. (2000). CpG DNA induces stronger immune responses with less toxicity than other adjuvants. Vaccine, 18(17), 1755-1762.
[91] Davis H.L., Weeranta R., Waldschmidt T.J., Tygrett L., Schorr J., Krieg A.M., (1998). CpG is a potential enhancher of specific immunity in mice immunized with recombinant hepatitis B surface antigen. Journal of Immunology, 160(2), 870-876.

[92] Krieg A.M., Yi A.K., Joachim S., Heather D.L. (1998). The role of CpG dinucleotides in DNA vaccines. Trends in Microbiology, 6(1), 23-27.

[93] Tang D.C., Devit M., Johnston S.A. (1992). Genetic immunization is a simple methods for eliciting an immune-response. Nature, 358(6365), 152-154.

[94] Redding L., Weiner D.B. (2009). DNA vaccines in veterinary use. Expert Review of vaccines, 8(9), 1251-1276.

[95] Hong W., Xiao S., Zhou R., Fang R., He Q., Wu B., Zhou F., Chen H. (2002). Protection induced by intramuscular immunization with DNA vaccines of pseudorabies in mice, rabbits and piglets. Vaccine, 20(7-8), 1205-1214.

[96] Feltquate D.M., Heaney S., Webster R.G. (1997). Different T helper cell types and antibody isotype generated by saline and gene gun DNA immunization. Journal of Immunology, 158(5), 2278-2284.

[97] Pertmer T.M., Roberts T.R., Haynes J.R. (1996). Influenza virus nucleoprotein specific immunoglobulin G subclass and cytokine responses elicited by DNA vaccination are dependent on the route of vector DNA delivery. Journal of Virology, 70(9), 6119-6125.

[98] Babiuk S., Beca-Estrada M., Foldvari M., Storm M., Rabussay D., Widera G., Babiuk L.A. (2002). Electroporation improves the efficacy of DNA vaccines in large animals. Vaccine, 20,(27-28), 2299-3408.

[99] Watts A.M., Kennedy R.C. ( 1999). DNA vaccination strategies against infectious diseases. International Journal for Parasitology, 29, 1149-1163.

In: DNA Vaccines: Types, Advantages and Limitations
Editors: E. C. Donnelly and A. M. Dixon
ISBN 978-1-61324-444-9


*Chapter IX*

# DNA Vaccination: Progress and Challenges

***Marcelo Sousa Silva,* Andreia Sofia Cruz Lança, Karina Pires de Sousa and Jorge Atouguia***
Centre for Malaria and Tropical Disease, Instituto de Higiene e Medicina Tropical
Universidade Nova de Lisboa. Lisboa, Portugal

## Abstract

DNA vaccination is one of the most promising techniques for immunization against diseases caused by viruses, protozoa, bacteria, and even for tumours and illnesses with genetic origins. These vaccines can be administered by direct inoculation of plasmid by several routes. The DNA immunization can induce both humoral and cellular immune responses, directed to the CD4 helper T cells and CD8 cytotoxic T responses, production and presentation of antigen to the immune system in a manner similar to that of a natural infection, enabling lasting immune responses and the possibility of combining a few antigens for the simultaneous treatment of many diseases. Another major advantage of DNA vaccines is the low toxicity, as DNA is completely innocuous. The use of DNA vaccines offers a number of economic, technical and logistics, advantages as compared to classical vaccines. Its mass production is cheaper, allows for easier control of the quality, and marketing does not require a cold chain, because these vaccines are stable at room temperature and endure pH variations. These factors facilitate the transport and distribution, and allow the transfer of technology. Despite the potential of DNA vaccines is not always fully achieved, since it depends on the nature of antigens, frequency and route of administration, the concentration of DNA administered targets the cellular localization of the antigen encoded by plasmid regardless of host age and health. Regarding the security of this new technology, one must take into account the possibility of plasmid integration into the host genome as potentially damaging, as exposure to high levels of antigenic proteins of the immune system during the period of gene expression may lead to generation of autoimmune diseases. In short, although this methodology still

* Corresponding Author: Instituto de Higiene e Medicina Tropical, Rua da Junqueira, 100,1349-008 Lisboa, Portugal, Phone: (+351) 213652629, FAX: (+351) 213622458, E-mail: mssilva@ihmt.unl.pt.

has some way to go before it can be safely administered in humans or animals to cure disease; this method allows the development of scientifically progress and maturity.

## Abbreviations

| | |
|---|---|
| pDNA | plasmid DNA |
| APCs | antigen presentation cells |
| DC | dendritic cells |
| CTLs | cytotoxic T lymphocytes |

## Introduction

Vaccination is the most efficient and less expensive method to prevent infectious diseases. The first generation vaccines were produced with attenuated or live microorganisms, or with killed or inactivated microorganisms, able to induce T-cells and antibody responses [1]. In order to minimise the risks of the first vaccines, the so-called second-generation vaccines were developed, consisting in purified antigens from natural and synthetic sources and recombinant antigens. The evolution of molecular biology and the optimization of pathways of administration allowed improved safety, efficiency and versatility of such sub-unit vaccines. The third generation vaccines, also called gene vaccines or DNA vaccines, have emerged after the introduction of genes or gene fragments that encode for potentially immunogenic antigens from a pathogen into viral vectors or plasmid DNA (pDNA).

DNA vaccines are basically composed of pDNA encoding the vaccine antigens and saline solutions. The pDNA contain generally of four units: (1) an antibiotic resistance gene controlled by a prokaryotic promoter, (2) a prokaryotic origin of replication that allow for selection and replication of pDNA in transformed bacteria, (3) a strong and ubiquitous promoter that confers optimal expression of the gene of interest in eukaryotic cells, and (4) a termination-polyadenylation sequence. Additionally, pDNA contains unmethylated CpG sequences that lead to an adjuvant effect through the activation of the innate immune response [2]. Thus, DNA vaccines are injected into the host and use its own cellular machinery to read the gene and convert it to immunogenic protein. This protein is processed by the host cells, displayed on the cell surface and recognized as foreign by the immune system, triggering a wide range of immune responses [3].

In 1960 it was shown that papilloma was induced by the injection of nucleic acid from the Shoppe rabbit papilloma virus [4]. In 1962, studies demonstrated that polyoma virus DNA from infected cells injected into hamsters lead to anti-polyoma antibodies and tumors [5]. However, the real potential of these observations went through unnoticed until Wolff and colleagues [6] demonstrated that expression plasmids could effectively be transferred into muscle cells by injecting them as a simple saline solution. This work has propelled further studies on this technology and two years later, Tang and colleagues [3], already using the "gene gun" as a new delivery technique, propagated genetic immunization as a simple method for eliciting an immune response. One year later, the first reports on immunoprotective DNA vaccines in mice were published, using the influenza virus model and genes encoding the viral nucleoprotein [6] or hemagglutinin [7]. Thus, very rapidly, convincing evidence for a reliable concept had been produced and has encouraged research groups all over the world to build upon the foundation laid and triggered an explosion in the field of nucleic acid vaccination, with literally thousands of patents being filed in a few years.

The initial observations on DNA vaccines led to a series of studies intended to determine how such vaccines could work. These studies covered three general areas: (1) the source of antigen presentation, (2) the immunological properties of the DNA itself, and (3) the role of cytokines in eliciting the immune responses [8]. Today, it is known that there are at least three mechanisms by which DNA vaccines are processed and presented *in vivo* to elicit immune responses: (1) direct priming by somatic cells (e.g. myocytes and keratinocytes); (2) direct transfection of professional APCs (i.e. DCs); and (3) 'cross-priming', in which plasmid DNA transfects a somatic cell and/or professional APC and the secreted protein is processed by untransfected DCs and presented to T cells [9].

About the action mechanism of DNA vaccines, several studies with reporter genes showed that the method of delivery of the DNA affected the range of cell types that were transfected. Bombardment of the epidermis with plasmid coated onto gold microbeads tended to directly transfect epidermal keratinocytes and Langerhans cells, which were shown to migrate rapidly to regional lymph nodes [10]. In this case, the source of antigen presentation and co-stimulatory molecules appeared straightforward, because professional APCs were transfected directly. On the other hand, intramuscular injection of plasmid predominantly led to transfection of myocytes [11]. Direct uptake of plasmid by professional APCs after intramuscular injection was much more difficult to demonstrate directly and appeared to be much less frequent [12]. Nonetheless, bone marrow-derived APCs were shown to be absolutely required for the induction of MHC class 1-restricted CTLs after intramuscular DNA vaccination [11]. However, further studies are needed to clarify even further the mechanisms of DNA vaccination and the induction of the immune system.

Nowadays, it has been successfully proven that it is possible to induce an immune response by direct inoculation of pDNA vaccines by several routes (intratracheal, intravenous, intrabursal, intraorbital, intradermal, intramuscular, oral, intranasal, intratumour, subcutaneous and mucosal route, etc.), keeping in mind that some pathways provide higher or lower level of antigen expression. These levels of expression appear to correlate with the amount of transfected cells obtained after administration of pDNA [13], and a potential also exists for variations in immune response due to differences in the duration of expression, or persistence of the antigen in somatic cells in the periphery or in APCs in lymphoid tissue [11].

In addition, is important to note that the amount of antigen produced *in vivo* after DNA inoculation is usually in the picogram to nanogram range. Given the relatively small amounts of protein synthesized by DNA vaccination, the efficient induction of immune responses must relate to the type of APCs transfected and/or the immune-enhancing properties of the DNA itself (i.e. CpG motifs) [14].

In summary, much must be considered regarding the success of immunization with DNA. It depends on the nature of the antigen, the frequency and route of administration, the concentration of DNA, the cellular location of the antigen encoded by the pDNA, the age and health of the host and the species of the animals vaccinated.

# Advantages of DNA Vaccines

In the earlier years of the 90s, the potential of a gene therapy designed to generate both cytotoxic and antibody immune responses was found to have many potential advantages over other vaccine strategies. Since then, a literal explosion in the field of nucleic acid vaccination has risen, with literally thousands of patents being filed in a few years [15].

Looking from a global point of view and considering that vaccination is the main system of preventing and/or reducing the impact of many pandemics, being able to make enough doses in time will hinge around the trade off between the capital investment required to have spare vaccine production capacity lying dormant for long periods of time versus the health benefits of having large numbers of vaccine doses when required to combat a pandemic or epidemic [16]. Furthermore, the cost of production of DNA vaccines, when compared to any other type of vaccine, is also much lower [17]. Hence, it will be cheaper to mass-produce genetic vaccines and this is an important consideration for developing countries or in veterinary medicine. Furthermore, they are extremely stable; DNA can be boiled, precipitated in ethanol and shipped across the globe at room temperature, without the requirement of cold chains, which are very difficult to maintain in less-developed countries [18].

One major point that must be enlisted on the advantages of the DNA vaccines is the ease with which they can be generated, modified and purified. Whilst generating and producing a purified an attenuated vaccine might several months, generation of a DNA vaccine takes only a few weeks [18]. As different DNA molecules have such similar physicochemical characteristics and production methods can be near-duplicated for different molecules, this time could conceivably be reduced to around 2 weeks [16].

In general, the major advantage of DNA vaccines is that the product is plasmid DNA, regardless of the antigen gene or disease target. Therefore, development costs for new DNA vaccines should be minimal once generic procedures are developed. In addition, the practicality of developing combination vaccines consisting of multivalent pDNA (bicistronic vector) or multiple plasmids expressing different antigens should be straightforward from the perspective of formulation since all components are DNA [19].

Another advantage of DNA vaccines is that they are not infectious, but they produce their antigens *in vivo*, which is an important pre-requisite for stimulating strong cytotoxic T-cells and optimal immune responses [20, 21]. Also, making the vaccine's own cellular machinery responsible for producing the immunogenic protein assures optimal glycosylation and folding of the antigen [4, 15]. Therefore, there is a warrant of antigen presentation with native posttranslational modifications, conformation, and oligomerization to elicit antibodies of optimal specificity that could furthermore provide long-lasting immune responses.

One of the advantages of DNA vaccines is the fact that a non-viral vector is used for immunization. Although viral-mediated gene transfer has a high transfection efficiency and stability, the largest hurdles using viral vectors are to overcome the immunogenicity of the viral packaging proteins [22]. Also, non-viral vectors are highly flexible, are capable of encoding a number of immunological components, are associated with a lower cytotoxicity, are relatively more stable, and are potentially more cost-effective for manufacture and storage [22]. Furthermore, it shows very little dissemination and transfection at distant sites following delivery and can be re-administered multiple times into mammals (including primates) without inducing an antibody response against itself (i.e., no anti-DNA antibodies generated)

[23]. Also, contrary to common belief, long-term foreign gene expression from naked pDNA is possible even without chromosome integration if the target cell is post mitotic (as in muscle) or slowly mitotic (as in hepatocytes) and if an immune reaction against the foreign protein is not generated [22, 24, 25]. In addition, with the advent of intravascular and electroporation techniques, its major restriction – poor expression levels – is no longer limiting and levels of foreign gene expression *in vivo* are approaching what can be achieved with viral vectors [26].

Vaccination has provided a very cost-effective approach to prevent some infectious diseases, but conventional vaccines display some limitations in the early age group: poor immunogenicity in case of killed or subunit vaccines, potential side effects for live vaccines or inhibition by maternal antibodies in all these circumstances [27]. In this context, DNA vaccines have also been reported as capable of overcoming maternal immunity [28, 29, 30] which is particularly important since the vertical transmission of several diseases is a recurrent event in many countries and is responsible for the high toll of death among first-week infants.

Also, the induction of tolerance or a state of no responsiveness was previously thought to preclude vaccination as an effective therapy in the fetus or newborn [31]. Increasing evidence, however, indicates that fetal DNA immunization can induce active immunity in the newborn. Early neonatal immune protection may find an important and life-saving application in hereditary or congenital diseases. The induction of protective immune responses *in utero* could have a significant impact on survival and quality of life for the large number of infants infected during or shortly after birth [29]. On this matter, DNA-based vaccination technology is at the frontline of the therapeutic efforts; and if efficient and consistent oral DNA delivery can be achieved *in utero*, then this route of delivery might provide new strategies for both vaccination and gene therapy [29].

Another of the noteworthy advantages of DNA-based drugs over currently available pharmaceuticals is their selective recognition of molecular targets and pathways, which imparts tremendous specificity [32]. Therefore, DNA vaccines could be used to mitigate disease (even prophylactically), thereby preventing disease progression and its complications. This technology permits the correction of a malfunctioning gene by the introduction and expression of its correct copy, thus resulting in a single protein product, thus ensuring specificity in controlling the disease status.

Moreover, the DNA immunization offers the potential combination of diverse immunogens into a single preparation, which facilitates simultaneous immunization for several diseases. They can also be used repeatedly for different immunogens [18]. In summary, the DNA vaccinations presents several advantages (listed in the figure 1) when compared to more traditional vaccination types, and are sure to be in the next years in the vanguard of therapeutical efforts.

## Limitations of DNA Vaccines

After some 15 years of experimentation, DNA vaccines have become well established as a research tool in animal models. However, DNA vaccines so far have shown low immunogenicity when tested alone in human clinical trials.

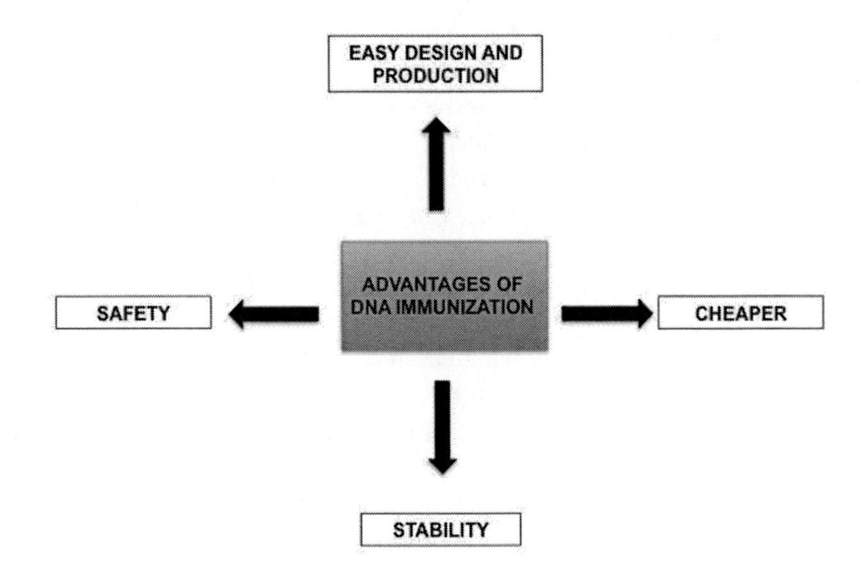


Figure 1. Advantages of DNA vaccines approach.

A significant effort has been put forward to identify methods of enhancing the immune response to plasmid DNA to enable its general use as a method of immunization in humans. So far, the improvements that have been seen are incremental, but this work is both continuing and making progress. The knowledge that is being gained in the pursuit of more effective DNA vaccines also is enriching the development of "conventional" vaccine approaches, and this understanding may well facilitate the invention of effective new vaccines for cancer and infectious diseases [11].

Although DNA immunization can elicit strong immune responses, the results obtained in non-human primates and humans revealed weak immune responses when compared to the ones observed in other animal models. Thus, DNA vaccines are considered less potent that the "traditional" live attenuated vaccines, and one of the reasons for this is that the DNA is not distributed evenly between the cells initially transfected, while the number of cells infected by the attenuated microorganism increases when it replicates. Also, the uptake of plasmid by the cells after injection is inefficient because only a small part of the injected material is transfected successfully: the intramuscular and intradermal release of plasmid DNA causes its rapid degradation by nuclease action in the extracellular media. To compensate for this, different administration routes must be considered and tested, since the effectiveness of these genetic vaccines is influenced by various factors, such as the (1) size of vector, (2) adsorption to particles, (3) host species and (4) the cellular localization of the expressed antigen.

A different strategy to increase the efficiency of DNA vaccines is the use of adjutants that increase the immunogenicity of the antigen or pDNA formulations. Additionally, the use of the gene gun system is also a mean to compensate the difficulty in delivering the DNA vaccine to the APCs and preventing the loss of injected material by low transfection rates due to nuclease action, since it allows the release of DNA inside the cells. Also, while intramuscular immunization preferentially induces the Th1 response, the gene gun system used on the dermis stimulates a Th2 response or a balanced Th1/Th2 response, which characterizes the humoral immune response [18].

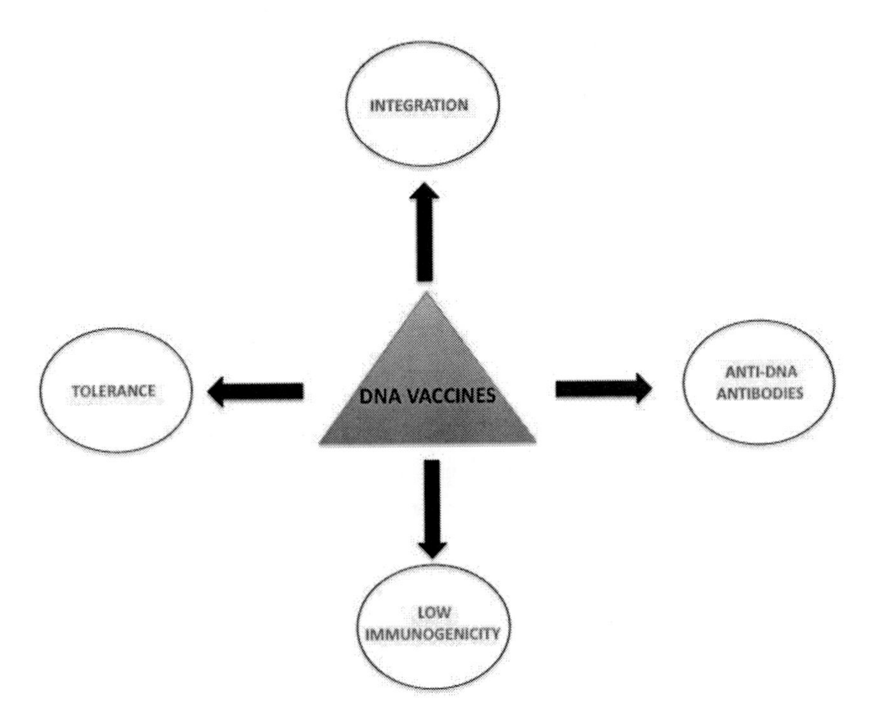


Figure 2. Potential concerns associated with DNA immunization.

In humans, there is an increased difficulty in inducing a strong immune response by immunization with genetic vaccines due to the low production of antigens, the cellular release of plasmid DNA and the inefficient stimulation of the innate immune system. There have been no changes in the vectors to improve the potency of the vaccine by increasing the efficiency of the promoter, but instead, two other physical strategies were used considering the great stability of the plasmid vaccine in the host cell (due mainly to their small size and negative charge that causes repulsion). One such strategy is the use of *in situ* electroporation or the direct tattooing of DNA into skin cells. The other strategy is a formulation with micro particles to target APCs, on the premises that the genetic material can be coated with biodegradable microspheres that protect and gradually allow the release of the DNA, which is rapidly phagocytised by macrophages and dendritic cells due to its small size. For the coating of the genetic vaccine, liposome and viral particles can be used too.

There are several features about the use of DNA vaccines that raise hypothetical concerns, which, in the light of inexperience of their use, have to be addressed at the preclinical safety stage. These concerns are represented in the figure 2.

## Conclusion

The term "gene therapy" has evolved into the so-called DNA vaccination, which nowadays represents one of the most notable tools under development in the field of vaccinology. Although the implementation of this approach for vaccination should overcome several bottlenecks, the advantages of this technique make it worth the efforts that are being invested in the establishment of these third-generation vaccines in the field. The use of pDNA represents an important platform for clinical applications, in which large-scale vaccine production is not easily manageable with other forms of vaccine including recombinant protein, whole tumour cells, or viral vectors.

The great advantage of this type of vaccine is the ability to cause an immune response similar to that of a natural infection even though it doesn't contain the microorganisms, only copies of some of its genes. Therefore, gene vaccines take the process of immunization to a new technological level, also being relatively easy and inexpensive to design and produce.

The remarkable numbers of publications demonstrating the efficacy of DNA vaccines in various preclinical models that have appeared since the publication of the initial demonstration of the generation of protective efficacy attest to the simplicity as well as the robustness of the technology. Indeed, in addition to viral disease models, preclinical efficacy or immune responses have been shown for models of bacterial and parasitic diseases, allergy, tumors, and other antigens [9]. Research in terms of human medicine has been directed mainly to HIV, malaria and tuberculosis as well as controlling the growth of melanoma in advanced stages.

## Acknowledgments

The authors thank the Portuguese Fundação Calouste Gulbenkian for fund support and Fundação para a Ciência e a Tecnologia for fund support (PTDC/CVT/72624/2006; PTDC/CVT/102486/2008) and post-doctoral grants to Silva MS (SFRH/BPD/26491/2006).

## References

[1] Graham, A., Timothy F. Brewer, Catherine S. Berkey,, Mary E. Wilson, Elisabeth Burdick, Harvey V. Fineberg, Frederick Mosteller. Efficacy of BCG vaccine in the prevention of tuberculosis. *JAMA*, 271(9): 698-702, 1994.

[2] Sato Y, Roman M, Tighe H, Lee D, Corr M, Nguyen M-D *et al*. Imunostimulatory DNA sequences necessary for effective intradermal gene immunization. *Science,* 273: 352-354, 1996.

[3] Tang, D., Devit, M., Johnston, S.A. and others. Genetic immunization is a simple method for eliciting an immune response. *Nature,* 356 (6365): 152–154, 1992.

[4] Wolff L and R Koller. Regions of the Moloney murine leukemia virus genome specifically related to induction of promonocytic tumors *J. Virol.,* 64(1): 155-160, 1990.

[5] Ito, Yohei. Relationship of components of papilloma virus to papilloma and carcinoma cells. *Cold Spring Harb. Symp. Quant. Biol*. 27: 387-39. 1962.

[6] Atanasiu P, Cannon Da, Dean Dj, Fox Jp, Habel K, Kaplan Mm, Kissling Re, Koprowski H, Lepine P, Perez Gallardo F. Rabies neutralizing antibody response to different schedules of serum and vaccine inoculations in non-exposed persons.

[7] Ulmer JB, Donnely JJ, Parker SE, Rhodes GH, Felgner PL, Dwarki VJ, Gromkowski SH, Deck RR, De Witt CM, Friedman A, Hawe LA, Leander KR,

Martinez D, Perry HC, Shiver JW, Montgomery DL e Liu MA. Heterologous protection against Influenza by injection of DNA encoding a viral protein. *Science,* 259: 1745-1749, 1993.

[8] Fynan E.F., Webster RG, Fuller DH, Haynes JR, Santoro JC, and Robinson HL. DNA vaccines: protective immunizations by parenteral, mucosal, and gene-gun inoculations. *Proc. Natl. Acad. Sci. USA*, 90(24): 11478–11482, 1993.

[9] Donnelly JJ, Ulmer JB, Shiver JW e Liu MA. DNA vaccines. *Annu. Rev. Immunol.*, 15: 617-648, 1997.

[10] Gurunathan S, Sacks DL, Brown DR, Reiner SL, Charest H, Glaichenhaus N e Seder RA. Vaccination with DNA encoding the immunodominant LACK parasite antigen confers protective immunity to mice infected with Leishmania major. *J. Exp. Med.*, 186: 1137-1147, 1997.

[11] Donnelly JJ, Liu MA e Ulmer JB. Antigen presentation and DNA vaccines. Am. J. Respir. Crit. *Care Med.*, 162: 190-193, 2000.

[12] Porgador A, Irvine KR, Iwasaki A, Barber BH, Restifo NP, and Germain RN. Predominant Role for Directly Transfected Dendritic Cells in Antigen Presentation to CD8+ T Cells after Gene Gun Immunization. *Exp Med.*, 188(6): 1075–1082, 1998.

[13] Lu S., Wang S., and Grimes-Serrano, JM. Current progress of DNA vaccine studies in humans. Expert Rev. *Vacccines*, 7:2, 175–191, 2008.

[14] Gurunathan S, Prussin C, Sacks DL, Seder RA. Vaccine requirements for sustained cellular immunity to an intracellular parasitic infection. *Nat. Med.,* 4(12): 1409-1415, 1998.

[15] Becker PD, Noerder M, Guzmán CA. Genetic immunization – bacteria as DNA vaccine delivery vehicles. *Hum. Vac.,* 4(3): 189-202, 2008.

[16] Han Y, Liu S, Ho J, Danquah MK, Forde GM. Using DNA as a drug -bioprocessing and delivery strategies. *Chem. Eng. Res. Desig.*, 87: 343–348, 2009.

[17] Manoj S, Babiuk LA, van Drunen Little, van den Hurk S. Approaches to enhance the efficacy of DNA vaccines. *Critical Rev. Clin. Lab. Sci.,* 41(1):1-39, 2004.

[18] Kowalczyk DW, Ertl HCJ. Immune responses to DNA vaccines. *Cellul. Mol. Life Scie.*, 55: 751–770, 1999.

[19] Montgomery DL, Ulmer JE, Donnelly JJ, and Liu MA. DNA Vaccines. *Pharmacol. Ther.,* 74:2: 195-205, 1997.

[20] Gregersen JP. DNA vaccines. *Naturwissenschaften,* 88:504–513, 2001.

[21] Lowe DB, Shearer MH, Jumper CA, Kennedy RC. Towards progress on DNA vaccines for cancer. *Cel. Mol. Life Sci.,* 64: 2391-2403, 2007.

[22] Fioretti D, Iurescia S, Fazio VM, Rinaldi M. DNA Vaccines: Developing New Strategies against Cancer. *J. Biom. Biotechn.*, 2010: 3-16, 2010.

[23] Jiao S, Williams P, Berg RK, Hodgeman BA, Liu L, Repetto G, Wolff JA. Direct gene transfer into nonhuman primate myofibers in vivo. *Hum. Gene Ther.*, 3(1): 21-33, 1992.

[24] Herweijer H, Zhang G, Subbotin VM, Budker V, Williams P, Wolff JA. Time course of gene expression after plasmid DNA gene transfer to the liver. *J. Gene Med.*, 3(3): 280-291, 2001.

[25] Wolff JA, Ludtke JJ, Acsadi G, Williams P, Jani A. Long-term persistence of plasmid DNA and foreign gene expression in mouse muscle. *Hum. Mol. Gen.*, 1(6): 363-369, 1992.

[26] Wolff JA, Budker V. The mechanism of naked DNA uptake and expression. In: Non-Viral Vectors for Gene Therapy, Second Edition. *Advances in Genetics*, 54: 1-20, 2005.

[27] Bot A, Bona C. Genetic immunization of neonates. *Microb. Infect.*, 4: 511–520, 2002.

[28] Bot A. DNA vaccination and the immune responsiveness of neonates. *Intern. Rev. Immunol.*, 19(2-3): 221-45, 2000.

[29] Gerdts V, Snider M, Brownlie R, Babiuk LA, Griebel PJ. Oral DNA vaccination in utero induces mucosal immunity and immune memory in the neonate. *J. Immunol.,* 168: 1877-1885, 2002.

[30] Fazio VM, Ria F, Franco E, Rosati P, Cannelli G, Signori E, Parrella P, Zaratti L, Iannace E, Monego G, Blogna S, Fioretti D, Iurescia S, filippetti R, Rinaldi M. Immune response at birth, long-term immune memory and 2 years follow-up after in-utero anti-HBV DNA immunization. *Gene. Ther,* 11(6): 544-551, 2004.

[31] Butts C, Zubkoff I, Robbinsj DS, Cao S and Sarzotti M. DNA immunization of infants: potential and limitations. *Vaccine,* 16: 14115: 1444-1449, 1998.

[32] Patil SD, Rhodes DG, Burgess DJ. DNA-based therapeutics and DNA delivery systems: a comprehensive review. *Am. Assoc. Pharmac. Scien.* J., 7:1-9, 2005.

# Index

## B



## C

## D

## E

# I

## J



## K



## L

## M

## N



## O



## P

## Q



## R



## S

## T

## U



## V

## W



## Y